NEW INSIGHTS ON THE DEVELOPMENT OF THE VASCULAR SYSTEM

NEW INSIGHTS ON THE DEVELOPMENT OF THE VASCULAR SYSTEM

DOMENICO RIBATTI
Department of Basic Medical Sciences, Neurosciences, and Sensory Organs, University of Bari Medical School, Bari, Italy

ELSEVIER

ACADEMIC PRESS
An imprint of Elsevier

Academic Press is an imprint of Elsevier
125 London Wall, London EC2Y 5AS, United Kingdom
525 B Street, Suite 1650, San Diego, CA 92101, United States
50 Hampshire Street, 5th Floor, Cambridge, MA 02139, United States
The Boulevard, Langford Lane, Kidlington, Oxford OX5 1GB, United Kingdom

ISBN 978-0-323-90599-2

For information on all Academic Press publications
visit our website at https://www.elsevier.com/books-and-journals

Publisher: Stacy Masucci
Acquisitions Editor: Glyn Jones
Editorial Project Manager: Timothy Bennett
Production Project Manager: Omer Mukthar
Cover Designer: Matthew Limbert

Typeset by STRAIVE, India

Contents

Preface

The cardiovascular system plays a crucial role in vertebrate development and homeostasis. During embryonic development, the vasculature is formed by both vasculogenesis and angiogenesis. Vasculogenesis, which consists of de novo vessel formation from angioblasts, provides the nascent vascular network, particularly during early embryonic life. Angiogenesis, which refers to the expansion of a preexisting vascular bed through sprouting, bridging, and intussusceptive growth, intervenes mostly during the later stages of embryogenesis. Recent experimental studies have demonstrated that bone marrow-derived stem and endothelial progenitor cells can in principle contribute to tissue repair by induction of neoangiogenesis.

Several genetic and epigenetic mechanisms are involved in the early development of the vascular system, and there is extensive literature on the genetic background and the molecular mechanisms responsible for blood vessel formation. Evidence is now emerging that blood vessels themselves have the ability to provide instructive regulatory signals to the surrounding nonvascular target cells during organ development. Thus, endothelial cell signaling is currently believed to promote fundamental cues for cell fate specification, embryo patterning, organ differentiation, and postnatal tissue remodeling. The ability of vessels to influence the surrounding nonvascular cells may depend on the intrinsic heterogeneity of endothelial cells. Understanding the concept of vascular bed specificity represents a major challenge for future investigations. Indeed, one of the most interesting theoretical perspectives and practical applications of endothelial cell signaling is the possibility for these cells to maintain their inductive potential during adult life.

CHAPTER 1

Introduction

William Harvey established the modern functional definition of the circulatory system in the 17th century, describing how the arteries and veins carry blood away from and toward the heart in a continuous circulation (Fig. 1) (Ribatti, 2009). Harvey refuted the Galenic physiology of the formation of the blood in the liver and its consumption within organs. In his publication, *Exercitatio anatomica de motu cordis et sanguinis in animalibus* (Harvey, 1628), he postulated the existence of a circulatory movement of the blood (Harvey, 1963). He found support from Marcello Malpighi, who described capillaries in the lung and the frog mesentery (Brown and Barnes, 1994).

In the early 20th century, pioneering vascular embryologists, such as Florence Sabin, used careful examination of developing blood vessels to provide detailed anatomical descriptions of the formation of the earliest blood vessels in different mammalian species.

The cardiovascular system consists of the heart and an interconnected network of blood vessels, including the arteries, veins, and capillaries. The arteries and veins are classified on the basis of their size and location within the vascular tree.

The formation of the cardiovascular system is one of the earliest events to occur during embryonic development, and the vertebrate heart is the first organ to become functional and is controlled by both genetic and epigenetic (environmental) factors. The heart develops from the precardiac lateral folds derived from presumptive mesodermal cells to form the primitive heart tube, consisting of an inner endothelium, which is separated from the outer myocardial tube by the elastic cardiac jelly. Emergence of the cardiac endothelium and cardiomyocytes occurs almost concomitantly, and, at first, they develop rather independently from one another (Sugy and Markwald, 1996). The endocardium is continuous with the endothelium of the major blood vessels, the axial vein, and the dorsal aorta. Complex asymmetric morphogenetic movements coupled with uneven growth rates contribute to looping and chamber formation (Fig. 2).

The cardiovascular system has evolved to service the function of gas exchange, nutrient delivery, and waste product removal and contributes to

New Insights on the Development of the Vascular System
https://doi.org/10.1016/B978-0-323-90599-2.00001-5

Fig. 1 A portrait of William Harvey. *(Reproduced from Ribatti D: William Harvey and the discovery of circulation of the blood, J Angiogenes Res 1:3, 2009.)*

the control of temperature and pressure/perfusion through vasoconstriction and vasodilation. The principal function of the circulatory system remains to maintain oxygen homeostasis by modulating oxygen delivery within every tissue (Fig. 3). Several genetic, microenvironmental, and epigenetic mechanisms are involved in the early development of the vascular system.

Vascular development and homeostasis are underpinned by two fundamental features: the generation of new vessels to meet the metabolic demands of underperfused regions and the elimination of vessels that do not sustain flow. The generation of blood vessels was first described through careful observations of living amphibian or avian embryos (Clark, 1918; Sabin, 1917, 1920).

In 1878, Wilhelm Roux postulated that vascular development proceeds in three phases: a stage of primary differentiation founded on hereditary principles, a transitional stage wherein the genetic elements are gradually supplanted by functional adaptation, and a final stage wherein further vascular development is entirely regulated by mechanical forces (Roux, 1878).

Angiogenesis, describing the development of microvessels from preexisting ones, was recognized as early as 1794, when John Hunter in his *Treatise on the blood, inflammation and gunshot wounds* described the formation

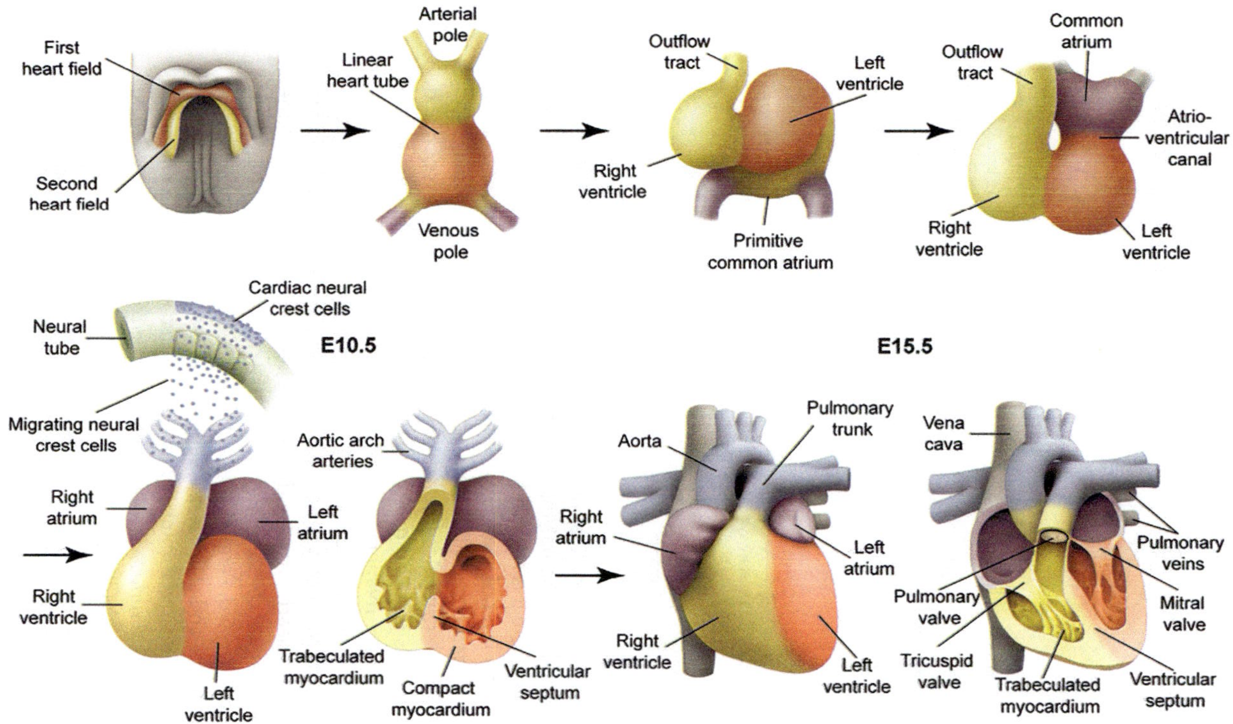

Fig. 2 The heart originates from mesodermal cells in the primitive streak. During gastrulation, cardiac progenitors migrate to the splanchnic mesoderm to form the cardiac crescent. At E7.5 in the mouse, the cardiac crescent can be divided into two heart field lineages based on differential gene expression and their respective contribution to the heart, namely, a first heart field *(red)* and a second heart field *(yellow)*, which are located posteriorly and medially to the first heart field. At E8.0, the linear heart tube is present. At E8.5, the looping is associated with uneven growth of cardiac chambers. The outflow tract is at the arterial pole, and the inflow tract and primitive atria are at the venous pole. By E9.5, the common atrium has moved superior to the ventricles and is separated by a distinct atrioventricular canal. By E10.5, cardiac neural crest cells from the dorsal neural tube migrate via the pharyngeal arches to the cardiac outflow tract. Further cardiac development involves a series of septation events and myocardial trabeculation that result in a mature four-chambered heart integrated with the circulatory system depicted at E15.5. *(Reproduced from Epstein JA, Aghajanian H, Singh MK: Semaphorin signaling in cardiovascular development, Cell Metab 21:163–173, 2015.)*

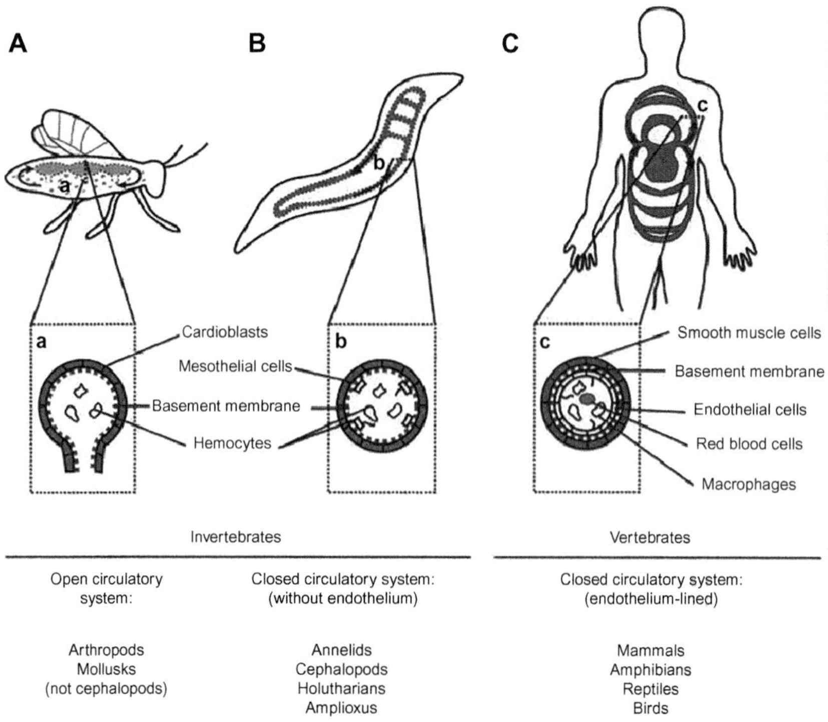

Fig. 3 Circulatory systems in multicellular organisms. (A) In arthropods and mollusks, the organs are surrounded by the hemolymph. The heart pumps the hemolymph into the open hemocoel from where it is retrieved back into the heart (arrows). (a) The vascular system is lined with a layer of cardioblasts and a luminal basement membrane. The hemolymph contains hemocytes. (B) In annelids, cephalopods, holothurians, and amphioxus, the blood is located in a closed vascular system. The latter consists of blood vessels in which the blood is pumped by contractile myoepithelial cells. (b) The vessels are lined by the basement membrane of mesothelial cells and/or intestinal cells. Hemocytes occur free in the vessel lumen and adherent at the luminal basement membrane. (C) In vertebrates, the circulatory system is closed. The blood is pumped through the whole organism by the heart. Arrows indicate the direction of blood flow. (c) The vasculature is lined by the apical cell surface of endothelial cells. The vascular basement membrane is abluminally located. Peripheral mural cells are located basally, within or at the basement membrane, and stabilize the vessels. The blood mainly contains red blood cells, besides leukocytes and thrombocytes. *(Reproduced from Strilic B, Eglinger J, Krieg GM, et al.: Electrostatic cell surface repulsion initiates lumen formation in developing blood vessels, Curr Biol 20: 2003–2009, 2010.)*

of vessels in the chick embryo (Fig. 4) (Hunter, 1794). He stated that vessels "would appear to have more powers of perfecting themselves, when injured, tan any other part of the body; for their use is almost immediate and constant, and it is they which perform the operation on the other parts, therefore they themselves must be perfect." (Hunter, 1861).

Fig. 4 A portrait of John Hunter. *(Reproduced from Wikipedia.)*

Most endothelial cells in adult organisms rarely divide. When they do, in response to physiological stimuli, such as wound healing and cycling in the female reproductive system, or to pathological stimuli, like in chronic inflammation or tumor growth, they participate in angiogenesis.

Blood vessels are believed to develop according to two fundamental processes. Vasculogenesis refers to embryonic formation of larger vessels, which arise in situ from primordial endothelial cells. Vasculogenesis predominates early in development and is responsible for initial assembly of vessels in the mammalian and avian yolk sacs and for formation of major intraembryonic blood vessels, such as the dorsal aorta and cardinal veins. The early vascular system consists of a network of immature endothelial channels. Conducting vessels (arteries and veins) develop later, following the metabolic demand of the organ primordia.

Environmental influences do not significantly determine initial vascular plexus generation, but hemodynamics and tissue oxygenation modulate early vasculogenesis.

The vascular network of certain endodermal organs, including the liver, lung, pancreas, stomach, intestine, and spleen, occurs by vasculogenesis. Otherwise, in the developing brain and kidney, the formation of the vascular tree occurs by angiogenesis.

Fibronectin is expressed in blood islands and in the primitive capillary plexus, and, in mice embryos lacking fibronectin, no blood vessels are formed in the yolk sac (George et al., 1993, 1997). Laminin, collagen, and other extracellular matrix molecules are synthesized later in vasculogenesis (Risau and Lemmon, 1988).

Vasculogenesis is not restricted to early embryogenesis, but contributes to newly forming adult blood vessels in physiological conditions, and, after myocardial and hind limb ischemia, to skin wounding, traumatic brain injury (Asahara et al., 1999; Guo et al., 2009; Kocher et al., 2001; Uribich and Dimmeler, 2004), and blood vessel formation in tumor growth in postnatal life, thorough incorporation of bone marrow-derived endothelial precursor cells (EPCs) (Ribatti et al., 2001).

Angiogenesis requires integrated activities of heterogeneous endothelial cell populations, and perturbations of any of these relationships compromise vessel development and function. More recent advances have introduced the concept of functional specialization of endothelial cells during the sprouting process in the so-called endothelial tip cells distinct from the endothelial stalk cells and have highlighted the important roles for components of the Notch and Wnt signaling in the regulation of sprouting angiogenesis. Proangiogenic signals, such as vascular endothelial growth factor (VEGF), promote endothelial motility, filopodia extension, and proliferation, and, under the control of Notch signaling, they determine whether endothelial cells become tip cells or stalk cells.

References

Asahara T, Masuda H, Takahashi T, et al.: Bone marrow origin of endothelial cells responsible for postnatal vasculogenesis in physiological and pathological neovascularization, *Circ Res* 85:221–228, 1999.

Brown A, Barnes J: William Harvey (1578-1657) and Marcello Malpighi (1628-1694): linked in blood, paralled in life, *Adler Mus Bull* 20:14–23, 1994.

Clark ER: Studies on the growth of blood vessels in the tail of the frog larva by observation and experiment on the living animal, *Am J Anat* 23:37–88, 1918.

George EL, Georges-Labouesse EN, Patel-King RS, et al.: Defects in mesoderm, neural tube and vascular development in mouse embryos lacking fibronectin, *Development* 119:1079–1091, 1993.

George EL, Baldwin HS, Hynes RO: Fibronectins are essential for heart and blood vessel morphogenesis but are dispensable for initial specification of precursor cells, *Blood* 90:3073–3081, 1997.

Guo X, Liu L, Zhang M, et al.: Correlation of $CD34^{+}$ cells with tissue angiogenesis after traumatic brain injury in a rat model, *J Neurotrauma* 26:1337–1344, 2009.

Harvey W: *Exercitatio anatomica de motu cordis et sanguinis in animalibus*, Farnkfurt, 1628, William Fitzer.

Harvey W: *The circulation of the blood and other writings*, London, 1963, Eveyman's Library.

Hunter J: *Treatise on the blood, inflammation and gunshot wounds*, London, 1794, George Nicol Bookseller.

Hunter J: *The works of John Hunter. Vol. 2. Observations in comparative anatomy*, London, 1861, John van Voorst.

Kocher AA, Schuster MD, Szabolcs MJ, et al.: Neovascularization of ischemic myocardium by human bone-marrow-derived angioblasts prevents cardiomyocyte apoptosis, reduced remodeling and improves cardiac function, *Nat Med* 7:430–436, 2001.

Ribatti D: William Harvey and the discovery of circulation of the blood, *J Angiogenes Res* 1:3, 2009.

Ribatti D, Vacca A, Nico B, et al.: Postnatal vasculogenesis, *Mech Dev* 100:157–163, 2001.

Risau W, Lemmon D: Changes in the vascular extracellular matrix during embryonic vasculogenesis and angiogenesis, *Dev Biol* 125:441–450, 1988.

Roux W: Uber die verzweigungen der Blutgefäβe des Menschen, *Jena Z Naturw* 12:205–206, 1878.

Sabin F: Origin and development of the primitive vessels of the chick embryo and of the pig, *Contrib Embryol* 18:62–124, 1917.

Sabin F: Studies on the origin of blood-vessels and of red-corpuscles as seen in the living blastoderm of chicks during the second day of incubation, *Contrib Embryol* 9:213–262, 1920.

Sugy Y, Markwald RR: Formation and early morphogenesis of endothelial precursor cells and role of endoderm, *Dev Biol* 175:66–83, 1996.

Uribich C, Dimmeler S: Endothelial progenitor cells: characterization and role in vascular biology, *Circ Res* 95:343–353, 2004.

CHAPTER 2
Vasculogenesis

Vasculogenesis is tightly controlled by genetic determinants and is characterized by cellular differentiation and decreasing plasticity. Until the onset of circulation, vascular development is genetically determined. Feedback signals from the vascular system, including pressure, flow, and metabolic state of the tissue, modify the genetically determined outline. Cardiac failure leads to early remodeling defects, and flow-induced eNOS expression controls secondary plexus generation (Lucitti et al., 2007).

Vasculogenesis leads to the formation of the first major intraembryonic blood vessels, such as the dorsal aorta and the cardinal veins, and to the formation of the primary vascular plexus in the yolk sac. Vasculogenesis begins extremely early after the initiation of gastrulation in the mammalian embryo, with the formation of blood islands in the yolk sac and precursor cells (angioblasts) in the head mesenchyme and the posterior lateral plate mesoderm.

The most typical and earliest site of prenatal vasculogenesis are the blood islands in the yolk sac, composed of hemangioblasts, the putative common precursors of endothelial and hematopoietic cells (Fig. 1) (Choi et al., 1998).

Blood islands develop as mesenchymal aggregates in the splanchnopleura of the area pellucida and the area opaca, collectively called the area vasculosa. Choi et al. (1998) reported that a single progenitor-derived embryoid body contains endothelial, hematopoietic, and smooth muscle lineages, implying the existence of hemangioblasts. In 2004, Huber et al. (2004) isolated potential hemangioblasts from E7.5 mouse embryos and showed that these cells give rise to both hematopoietic and vascular cells. Developing zebrafish gastrula contains progenitors that generate both endothelial and hematopoietic lineages (Vogeli et al., 2006).

Already in 1865, His (1865) described "angioblasts" as mesenchymal cells, which later form blood vessels. In 1917, Sabin (1917) adopted the term "angioblast" to describe these hypothetical progenitors. Sabin identified individual cells within the body of the embryo near the endoderm-mesoderm interface. These cells were preferentially distributed along the

New Insights on the Development of the Vascular System
https://doi.org/10.1016/B978-0-323-90599-2.00002-7

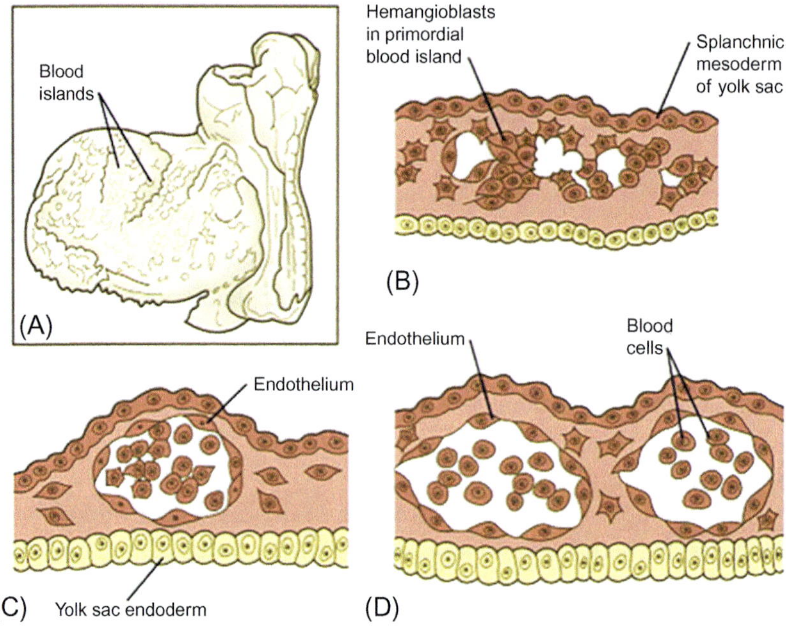

Fig. 1 Development of blood islands in the yolk sac of human embryos. (A) Gross view of a 10-somite human embryo showing the location of blood islands on the yolk sac. (B–D) Successive stages in the formation of blood islands. (Redrawn from Corner G: A well-preserved human embryo of 10 somites, *Contrib Embryol Carnegie* 20:81–102, 1929.)

paths of primary vessel formation. Many of them coalesce, forming elongated cords. Murray (1932) renamed these cells as "hemangioblasts," as he assumed that endothelial cells and blood cells would always develop from the same progenitor cell.

Cells from hematopoietic and endothelial lineages are first generated between the second and third week of human gestation (Oberlin et al., 2002). Hematopoiesis shifts to the fetal liver and again to the bone marrow (Moore and Metcalf, 1970). The majority of the inner cells of the blood islands progressively lose their intercellular attachments and differentiate into primitive erythroblasts.

Angioblasts, peripheral cells of the blood islands, connect to construct a primitive network composed of capillaries, arteries, and veins. Similarly, within the embryo itself, angioblasts start to join together to form a capillary network in the head mesenchyme and the posterior lateral plate mesoderm.

Interspecific grafting experiments using quail and chick embryos in combination with the QH1 antibody have demonstrated that the development of endothelial cells is independent of the formation of the mesoderm layer during gastrulation (von Kirschhofer et al., 1994).

When quail blastodiscs are explanted and disrupted in culture, mesoderm cells do not form angioblasts or express proteins associated with the endothelial cell lineage (Flamme and Risau, 1992), unless they are treated with fibroblast growth factor-2 (FGF-2),[a] in which case blood island-like structures are formed (Flamme and Risau, 1992). Flamme et al. (1997) showed that FGF-2 induces pluripotent cells of the quail blastodiscs to undergo vasculogenesis, and experiments in the chick have suggested that FGF signaling is important for initiation of angioblast specification (Cox and Poole, 2000). FGF-2 has been identified in the chick embryo chorioallantoic membrane (CAM) during vasculogenesis and can promote vessel growth when exogenously added to the CAM during embryo development (Fig. 2) (Ribatti et al., 1995). Moreover, neutralizing antibodies to FGF-2 inhibited vessel growth when applied locally (Fig. 2), suggesting that

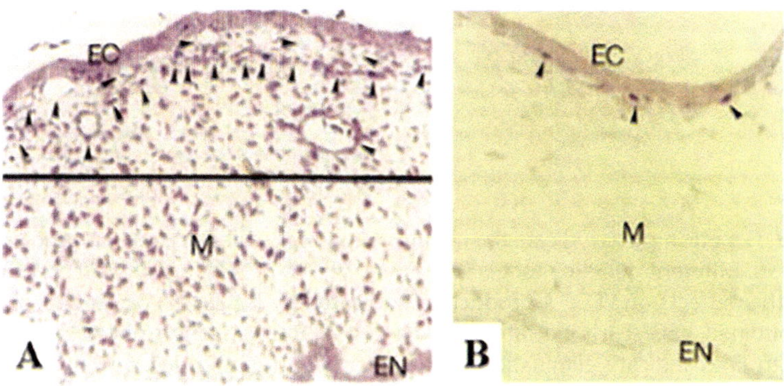

Fig. 2 Semithin sections of the CAM after treatment with FGF-2 (A) or anti-FGF-2 antibody (B). Note in A, an increased number of small blood vessels (arrowheads) beneath the ectoderm (EC) and in the upper portion of the intermediate mesenchyme (above the line); the mesenchyme (M) also contains numerous fibroblasts (EN, endoderm). Note in B, few vessels (arrowheads) beneath the ectoderm (EC), loosely arranged fibroblasts, and wide intercellular spaces; no blood vessels are detectable in the intermediate mesenchyme (M). (Reproduced from Ribatti D, Urbinati C, Nico B, et al.: Endogenous basic fibroblast growth factor is implicated in the vascularization of the chick chorioallantoic membrane, *Dev Biol* 170:39–49, 1995.)

[a] The FGF family comprises 23 different ligands and 4 tyrosine kinase-type receptors (FGFR-1–4). FGFR-1 and FGFR-2 are expressed by endothelial cells and their binding to ligands FGF-1, FGF-2, FGF-4, and FGF-5 leads to the induction of critical stages of angiogenesis in vivo.

FGF-2 normally functions to promote vessel growth, possibly by inducing angioblast formation from the mesoderm (Ribatti et al., 1995).

Both vascular endothelial growth factor (VEGF) and VEGF receptors (VEGFRs)[b] are the first endothelial-specific signal transduction pathways activated during vasculogenesis (Fig. 3). Once mesodermal precursors acquire the expression of VEGFR-2 and differentiate into angioblasts, they migrate away from the lateral plate mesoderm toward the VEGF gradient and coalesce as an endothelial cord of the presumptive dorsal aorta (Fish and Wythe, 2015). VEGF-deficient embryos die between days 8.5 and

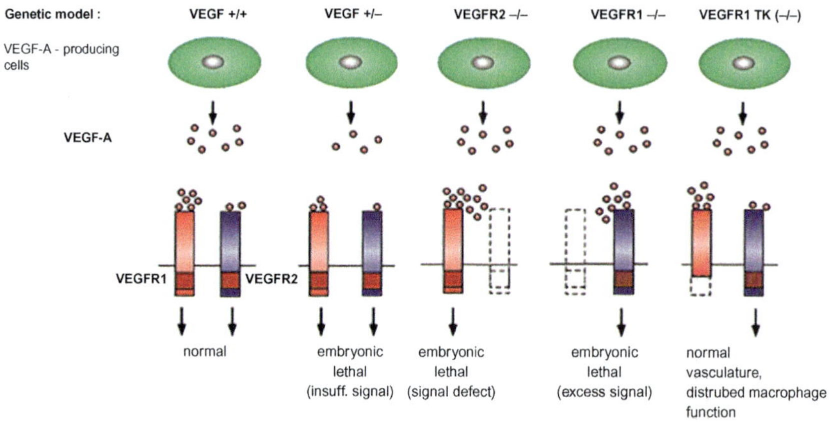

Fig. 3 Genetic models indicate the negative regulatory role of VEGFR-1 during embryogenesis. Different genetic models (VEGF +/+, VEGF +/–, etc. are indicated at the top) and their effects on VEGF-A production (indicated by *red circles*) or VEGFR expression are outlined. VEGFR-1 is indicated in *red* and VEGFR-2 is in *blue*. Unfilled symbols with broken lines indicate elimination by gene targeting. (Reproduced from Shibuya M, Claesson Welsh L: Signal transduction by VEGF receptors in regulation of angiogenesis and lymphangiogenesis, *Exp Cell Res* 312:549–560, 2006.)

[b] The VEGF family includes VEGF-A, VEGF-B, VEGF-C, VEGF-D, VEGF-E, and placental growth factor (PlGF). VEGF selectively binds to three receptors: VEGFR-1, VEGFR-2, and VEGFR-3. These receptors are principally expressed in endothelial cells. PlGF and VEGF-B bind to VEGFR-1, and VEGF-A can bind to VEGFR-1 and VEGFR-2. VEGF-C and VEGF-D bind to VEGFR-2 and VEGFR-3, and VEGF-E binds to VEGFR-2. Certain VEGF isoforms bind to non-tyrosine kinase coreceptors, neuropilin-1 and -2, to enhance VEGFR-2 activity. Five different isoforms of VEGF are generated by alternative splicing of a single gene: VEGF-121, VEGF-145, VEGF-165, VEGF-189, and VEGF-205. These isoforms differ in their binding affinity for heparin. All members of the VEGF family and their receptors work as homodimers although heterodimers between different members have also been reported.

9.0 and their primitive vascular structures are severely defective (Carmeliet et al., 1996a,b; Ferrara et al., 1996), whereas VEGFR-2-deficient mice die early because of the blocked migration of angioblasts to the initial sites of vasculogenesis (Shalaby et al., 1995, 1997). Embryos lacking VEGFR-2 die around 9 days of development and show no development of any blood vessels or hematopoietic cells (Shalaby et al., 1995). The loss of both lineages could suggest that VEGFR-2 is required in the common precursor, the hemangioblasts. Chimera studies suggest that the failure to form the hematopoietic lineage is a result of the incorrect migration and location of the progenitors of the endothelium and the blood system in the mutant yolk sac (Shalaby et al., 1997). VEGFR-2 is highly expressed later in endothelial cells (Millauer et al., 1993) but downregulated in most hematopoietic cells (Kabrun et al., 1997). In the yolk sac, VEGFR-2 expression appears to mark cells prior to the distinction between the outer endothelial cells and the enclosed hematopoietic cells (Yamaguchi et al., 1993), which led to the suggestion that it might also mark the hemangioblasts (Eichmann et al., 1997; Kabrun et al., 1997). Endothelial cells develop in VEGFR-1-deficient embryos, but death before day 9.5 still occurs due to the defective organization of the primitive vascular system (Fong et al., 1995). The phenotype is more suggestive of an overgrowth of endothelial cells than a loss of cells, as confirmed by chimera studies (Fong et al., 1999). This suggests that VEGFR-1 could be a negative regulator of VEGF signaling in the early embryo, acting to modulate VEGF by sequestering ligands. Inactivation of the gene for VEGFR-3 also affects lumen formation in large vessels during early development (Dumont et al., 1998). Early microinjection of VEGF into quail embryos caused vasculogenesis in normally avascular areas, while the normal polygonal pattern of vessel characteristics of early embryos was replaced by abnormal, enlarged vascular channels (Drake and Little, 1995).

The VEGF+/− embryos exhibited a number of developmental anomalies. The forebrain region appeared significantly underdeveloped. In the heart, the outflow region was grossly malformed; the dorsal aortas were rudimentary, and the thickness of the ventricular wall was markedly decreased. The yolk sac revealed a substantially reduced number of nucleated red blood cells within the blood islands, indicating that VEGF regulated both vasculogenesis and hematopoiesis. In addition, the vitelline veins failed to fuse within the vascular plexus of the yolk sac. Significant defects in the vasculature of other tissues, including the placenta and the nervous system, were evidenced. In the nervous systems of heterozygous embryos on day 10.5, vascular elements could be demonstrated in the mesenchyme, but not in the neuroepithelium, and the failure of blood vessel ingrowth was accompanied by apoptosis and disorganization of neuroepithelial cells.

The angiopoietin (Ang) family comprises at least four secreted proteins, namely, Ang-1, -2, -3 and -4, all of which bind to the endothelial-specific receptor tyrosine kinase Tie-2, whereas Tie-1 is an orphan receptor tyrosine kinase. It is well documented that Ang plays a critical role in endothelial sprouting, vessel wall remodeling, and mural cell recruitment (Thurston, 2003). Vasculogenesis proceeds normally in embryos lacking both Tie-1 and Tie-2 although they die early due to multiple cardiovascular defects (Suri et al., 1996; Puri et al., 1999). Mutation of Tie-2 does not affect initial formation of blood vessels, but embryos died in mid-gestation with major defects in vascular remodeling and stability (Dumont et al., 1994; Sato et al., 1995). Blood vessels were enlarged, with fewer branches, and endothelial cells tended to round up, dissociate from the underlying support cells and extracellular matrix, and undergo apoptosis. Chimera studies showed that the role of Tie-2 was specific to the late stages of capillary formation (Partanen et al., 1996).

Vasculogenesis takes place in a fibronectin-rich extracellular matrix (Risau and Lemmon, 1988). Fibronectin decreases in the vicinity of developing blood vessels, and endothelial cells begin to produce laminin and collagen type IV in increasing amounts (Risau and Lemmon, 1988; Drake et al., 1990; Ausprunk et al., 1991). Laminin expression in the vascular basement membrane is an early marker for vascular maturation. Fibronectin null mice die in utero at E9–E10 and exhibit severe defects in blood vessel and heart development (George et al., 1993).

Vasculogenesis can be demonstrated by scanning electron microscopy (De Ruiter et al., 1993). Endothelial precursors are located in the splanchnic mesoderm where they differentiate and delaminate from the mesothelium. All mesodermal tissues except the notochordal and prechordal plates contain angioblast precursors. Quail endothelial cells express a cell surface epitope, which is recognized by the QH1 antibody, which is also expressed by hematopoietic cells and endothelial precursors (Pardanaud et al., 1987).

The adult vasculature is a relatively quiescent organ, and the turnover of endothelial cells in healthy tissues is around 3 years (Cines et al., 1998). Vasculogenesis may not only be restricted to early embryogenesis but may also have a physiological role, thus contributing to the pathology of vascular diseases in postnatal life. The major regulator of postnatal vasculogenesis is VEGF. Springer et al. (1998) demonstrated that implantation into the mouse leg muscle of myoblasts transduced with VEGF led to a formation of new vessels similar to those formed during embryonic vasculogenesis. Localized constant production of high levels of VEGF can lead to recruitment of circulating endothelial precursor cells (EPCs) and induce them to undergo differentiation. Asahara et al. (1999) observed that VEGF-inducing mobilization of bone marrow-derived EPCs resulted in increased differentiated EPCs in vitro and augmented neovascularization in vivo. Postnatal vasculogenesis is governed by the same mechanisms as prenatal vasculogenesis. Tightly regulated vascular fusion by which small-caliber endothelial tubes fuse to enclose a common lumen to form larger vessels is likely a mechanism of vascular fusion (Drake and Little, 1999).

References

Asahara T, Masuda H, Takahashi T, et al.: Bone marrow origin of endothelial cells responsible for postnatal vasculogenesis in physiological and pathological neovascularization, *Circ Res* 85:221–228, 1999.

Ausprunk DH, Dethlefsen SM, Higgin ER: Distribution of fibronectin, laminin and type IV collagen during development of blood vessels in the chick chorioallantoic membrane, In *Issue in biomedicine*, 14, 1991, Karger, pp 93–108.

Carmeliet P, Ferreira V, Breier G, et al.: Abnormal blood vessel development and lethality in embryos lacking a single VEGF allele, *Nature* 380:435–439, 1996a.

Carmeliet P, Mackman N, Moons L, et al.: Role of tissue factor in embryonic blood vessel development, *Nature* 383:73–75, 1996b.

Choi K, Kennedy M, Kazarov P, et al.: A common precursor for hematopoietic and endothelial cells, *Development* 125:725–732, 1998.

Cines DB, Pollak ES, Buck CA, et al.: Endothelial cells in physiology and in pathophysiology of vascular disorders, *Blood* 91:3527–3561, 1998.

Cox CM, Poole TJ: Angioblast differentiation is influenced by the local environment: FGF-2 induces angioblasts and pattern vessel formation in the quail embryo, *Dev Dyn* 218:371–382, 2000.

De Ruiter MC, Poelman RE, Mentink MM, et al.: Early formation of the vascular system in quail embryos, *Anat Rec* 235:261–274, 1993.

Drake CJ, Little CD: Exogenous vascular endothelial growth factor induces malformed and hyperfused vessels during embryonic neovascularization, *Proc Natl Acad Sci U S A* 92:7657–7761, 1995.

Drake CJ, Little CD: VEGF and vascular fusion: implications for normal and pathological vessels, *J Histochem Cytochem* 47:1351–1355, 1999.

Drake CJ, Davis LA, Walters L, et al.: Avian vasculogenesis and the distribution of collagens I, IV, laminin, and fibronectin in the heart primordia, *J Exp Zool* 255:309–322, 1990.

Dumont DJ, Gradwohl G, Fong GH, et al.: Dominant-negative and targeted null mutations in the endothelial receptor tyrosine kinase, tek, reveal a critical role in vasculogenesis of the embryo, *Genes Dev* 8:1897–1909, 1994.

Dumont DJ, Jussila L, Taipale J, et al.: Cardiovascular failure in mouse embryos deficient in VEGF receptor-3, *Science* 282:946–949, 1998.

Eichmann A, Corbel C, Nataf V, et al.: Ligand-dependent development of the endothelial and hemopoietic lineages from embryonic mesodermal cells expressing vascular endothelial growth factor receptor, *Proc Natl Acad Sci U S A* 94:5141–5146, 1997.

Ferrara N, Carver-Moore K, Chen H, et al.: Heterozygous embryonic lethality induced by targeted inactivation of the VEGF gene, *Nature* 380:439–442, 1996.
Fish JE, Wythe JD: The molecular regulation of arteriovenous specification and maintenance, *Dev Dyn* 244:391–409, 2015.
Flamme I, Risau W: Induction of vasculogenesis and hematopoiesis in vitro, *Development* 116:435–439, 1992.
Flamme I, Frolich T, Risau W: Molecular mechanisms of vasculogenesis and embryonic angiogenesis, *J Cell Physiol* 173:206–210, 1997.
Fong GH, Rossant J, Gertsenstein M, et al.: Role of the Flt-1 receptor tyrosine kinase in regulating the assembly of vascular endothelium, *Nature* 376:66–70, 1995.
Fong GH, Zhang L, Bryce DM, et al.: Increased hemangioblast commitment, not vascular disorganization, is the primary defect in flt-1 knock-out mice, *Development* 126:3015–3025, 1999.
George EL, Georges-Labouesse EN, Patel-King RS, et al.: Defects in mesoderm, neural tube and vascular development in mouse embryos lacking fibronectin, *Development* 119:1079–1091, 1993.
His W: *Die Häute und Hölen des Körpers*, Basel, 1865, Schwighauser.
Huber TL, KouskpffV, Fehling HJ, et al.: Haemangioblasts commitments is initiated in the primitive streek of the mouse embryo, *Nature* 432:625–630, 2004.
Kabrun N, Buhring HJ, Choi K, et al.: Flk-1 expression defines a population of early embryonic hematopoietic precursors, *Development* 124:2039–2048, 1997.
Lucitti JL, Jones EA, Huang C, et al.: Vascular emodeling of the mouse yolk sac requires hemodynamic force, *Development* 134:3317–3326, 2007.
Millauer B, Wizigmann-Voos S, Schnürch H, et al.: High affinity VEGF binding and developmental expression suggest Flk-1 as a major regulator of vasculogenesis and angiogenesis, *Cell* 72:835–846, 1993.
Moore MA, Metcalf D: Ontogeny of the haemopoietic system: yolk sac origin of in vivo and in vitro colony forming cells in the developing mouse embryo, *Br J Haematol* 18:279–286, 1970.
Murray FP: The development in vitro of the blood of the early chicken embryo, *Proc R Soc Lond B* 111:497–521, 1932.
Oberlin E, Tavian M, Blazsek I, et al.: Blood-forming potential of vascular endothelium in the human embryo, *Development* 129:4147–4157, 2002.
Pardanaud L, Altmmann C, Kitos P, et al.: Vasculogenesis in the early quail blastodisc studied with monoclonal antibody recognizing endothelial cells, *Development* 100:339–349, 1987.
Partanen J, Puri MC, Schwartz L, et al.: Cell autonomous functions of the receptor tyrosine kinase TIE in a late phase of angiogenic capillary growth and endothelial cell survival during murine development, *Development* 122:3013–3021, 1996.
Puri MC, Partanen J, Rossant J, et al.: Interaction of the TEK and TIE receptor tyrosine kinases during cardiovascular development, *Development* 126:4569–4580, 1999.
Ribatti D, Urbinati C, Nico B, et al.: Endogenous basic fibroblast growth factor is implicated in the vascularization of the chick chorioallantoic membrane, *Dev Biol* 170:39–49, 1995.
Risau W, Lemmon D: Changes in the vascular extracellular matrix during embryonic vasculogenesis and angiogenesis, *Dev Biol* 125:441–450, 1988.
Sabin F: Preliminary note on the differentiation of angioblasts and the method by which they produce blood vessels, blood plasma, and red blood cells as seen in the living chick, *Anat Rec* 13:199–204, 1917.
Sato TN, Tozawa Y, Deutsch U, et al.: Distinct roles of the receptor tyrosine kinases tie-1 and tie-2 in blood vessel formation, *Nature* 376:70–74, 1995.
Shalaby F, Rossant J, Yamaguchi TP, et al.: Failure of blood-island formation and vasculogenesis in Flk-1-deficient mice, *Nature* 376:62–66, 1995.

Shalaby F, Ho J, Fisher KD, et al.: A requirement for Flk 1 in primitive and definitive hematopoiesis and vasculogenesis, *Cell* 89:981–990, 1997.

Springer ML, Chen AS, Kraft PE, et al.: BVEGF gene delivery to muscle: potential role for vasculogenesis in adult, *Mol Cell* 2:549–558, 1998.

Suri C, Jones PF, Patan S, et al.: Requisite role of angiopoietin-1, a ligand for the tie-2 receptor, during embryonic angiogenesis, *Cell* 87:1171–1180, 1996.

Thurston G: Role of angiopoietins and tie receptor tyrosine kinases in angiogenesis and lymphangiogenesis, *Cell Tissue Res* 314:61–68, 2003.

Vogeli KM, Jim SW, Martin GR, et al.: A common progenitor for haematopietic and endothelial lineages in the zebrafish gastrula, *Nature* 443:337–339, 2006.

von Kirschhofer K, Grim M, Christ B, et al.: Emergence of myogenic and endothelial cell lineages in avian embryos, *Dev Biol* 163:270–278, 1994.

Yamaguchi TP, Dumont DJ, Conlon RA, et al.: Flk-1, an flt-related receptor tyrosine kinase is an early marker for endothelial cell precursors, *Development* 118:489–498, 1993.

CHAPTER 3

Angiogenesis

Research in angiogenesis has greatly improved, thus allowing the characterization of different molecular mechanisms and the identification of several angiogenesis agonists and antagonists. Angiogenesis is defined as the sprouting of a new blood vessel from preexisting vessels. This can be accomplished through endothelial sprouting or nonsprouting (intussusceptive) microvascular growth (IMG) (Ribatti, 2006). The latter postulated that the capillary network increases its complexity and vascular surface by insertion of a multitude of transcapillary pillars, a process called "intussusception" (Djonov et al., 2000).

The first description of sprouting angiogenesis in tumor growth was reported by Ausprunk and Folkman (1977), which indicated the following stages: (1) the basement membrane is locally degraded on the side of the dilated peritumoral postcapillary venule situated close to the angiogenic stimulus; (2) the interendothelial contacts are weakened and endothelial cells migrate to the connective tissue; (3) a solid cord of endothelial cells forms; and (4) lumen formation occurs proximal to the migrating front and contiguous tubular sprouts anastomose to form functionally capillary loops, parallel to the synthesis of the new basement membrane and the recruitment of pericytes. Ausprunk and Folkman identified that blood vessel sprouting can initially progress without cell division, whereas sustained sprouting and further outgrowth require proliferation.

A variety of in vivo and in vitro models of angiogenesis have been developed, which have greatly contributed to the remarkable advances in this field. Among the in vivo assays, the three most used include the rabbit cornea (Muthukkaruppan and Auerbach, 1979), the developing mouse retina (Fruttiger, 2007), and the intersegmental vessel growth in zebrafish (Lawson and Weinstein, 2002). In vitro models with isolated endothelial cell lines have allowed the study of selected aspects of the angiogenic process, including endothelial migration, proliferation, proteolytic digestion of the extracellular matrix, and capillary tube formation (Cimpean et al., 2010).

Human endothelial cells were first placed on dishes coated with a thin layer of extracellular matrix proteins by Folkman and Haudenschild (1980)

New Insights on the Development of the Vascular System
https://doi.org/10.1016/B978-0-323-90599-2.00003-9

and organized into tube-like structures. Under both light and electron microscopes, Grant et al. (1991) showed that these capillary-like networks had lumens and that endothelial cells possessed membrane specialization similar to those observed in vessels in vivo. In vitro assays were further developed independently by Montesano and Madri (Montesano et al., 1983; Madri and Williams, 1983) using a three-dimensional collagen gel (Fig. 1). Montesano et al. (1983) demonstrated that when a monolayer of microvascular endothelial cells on the surface of the collagen gel is covered with a second layer of collagen, it reorganizes within a few days into a network of branching and anastomosing tubules.

In 1991, Paku and Pawoletz (1991) integrated the schema by Ausprunk and Folkman by means of ultrastructural observations, as follows: (1) a structural alteration of the basement membrane occurs, characterized by the loss of electron density over the entire circumference of the dilated mother vessel, followed by a partial degradation of the basement membrane at places where endothelial cell processes are projecting into the connective tissue;

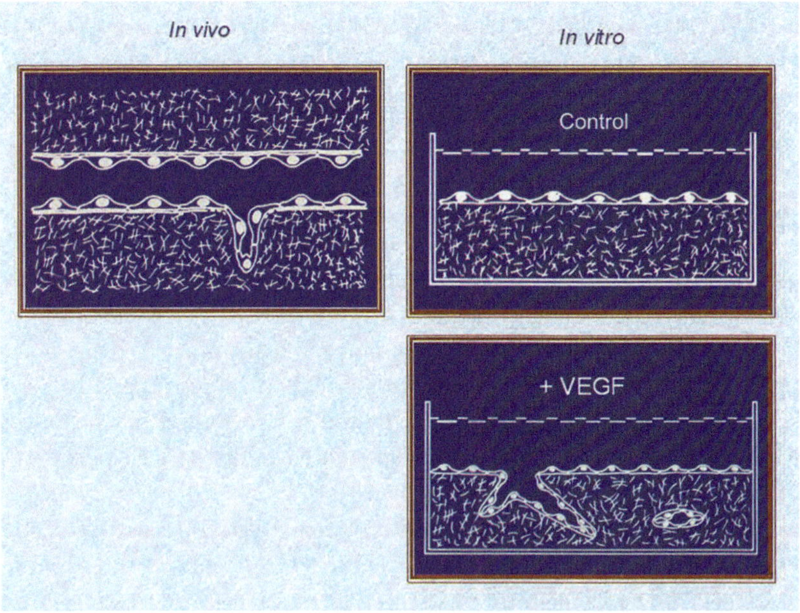

Fig. 1 Original paintings of Roberto Montesano, showing the three-dimensional cell culture system in which VEGF induces microvascular endothelial cells grown on collagen gels to invade the underlying matrix. (Reproduced from Ribatti D: The contribution of Roberto Montesano to the study of the interactions between epithelial sheets and the surrounding extracellular matrix, *Int J Dev Biol* 54:1–6, 2010a.)

(2) endothelial cells migrate, arranged in parallel, maintaining their basal-luminal polarity by forming a slit-like lumen and sealed by intact interendothelial junctions; (3) the basement membrane is deposited continuously by the polarized endothelial cells while only the tip of the growing capillary bud is devoid of the basement membrane; and (4) proliferating pericytes migrate along the basement membrane of the capillary bud, resulting in a complete coverage of the new vessel.

Werner Risau (Fig. 2) and co-workers found that the spatial and temporal expression patterns of vascular endothelial growth factor (VEGF) mRNA correlated with angiogenesis during embryonic development in the mouse brain (Breier et al., 1992). VEGF mRNA was expressed in the ventricular layer of the developing neuroectoderm, and the concentration gradient of the diffusible VEGF isoforms declining from the ventricular layer toward the perineural vascular plexus caused the ingrowth of capillaries from the vascular plexus toward the angiogenic stimulus provided by VEGF-secreting ventricular epithelial cells (Breier et al., 1992).

Flk-1/VEGF receptor-2 (VEGFR-2) is the first endothelial receptor tyrosine kinase known to be expressed on endothelial cell precursors and plays a central role in regulating embryonic vascular development and tumor angiogenesis. Lineage-tracing experiments demonstrated the presence at the ventral margin of the embryo at 1 h post gastrulation of a labeled population consisting exclusively of hematopoietic (gata-1-positive) and

Fig. 2 A portrait of Werner Risau. (Reproduced from Ribatti D: The seminal work of Werner Risau in the study of the development of the vascular system, *Int J Dev Biol* 54:567–572, 2010b.)

endothelial (flk-1/VEGFR-2-positive) lineages (Vogeli et al., 2006). Mice deficient in flk-1/VEGFR-2 died in utero between days 8.5 and 9.5, as a result of an early defect in the development of hematopoietic and endothelial cells and a complete lack of vasculature (Shalaby et al., 1995). These mice failed to develop yolk sac vasculature, indicating that flk-1/VEGFR-2 plays an essential role in survival/growth and differentiation of endothelial cell progenitors.

Risau and co-workers isolated flk-1/VEGFR-2 from proliferating endothelial cells of a postnatal mouse brain, when there was maximal endothelial cell proliferation (Millauer et al., 1993). This VEGF receptor was restricted to the capillaries and blood vessels, and when the first vascular sprouts began to radially invade the neuroectoderm from the perineural plexus, the expression of flk-1/VEGFR-2 was high in the perineural vascular plexus and in the invading vascular sprouts (Millauer et al., 1993). In the adult brain, when angiogenesis had ceased, the flk-1/VEGFR-2 expression was extremely low and restricted to choroid plexus endothelial cells (Kremer et al., 1997). Risau and co-workers (Kabrun et al., 1997) demonstrated that flk-1/VEGFR-2 provides a marker for the earliest detectable hematopoietic precursors. An RNA analysis indicated that it was expressed in the yolk sac on days 10–12 of gestation, in the liver throughout fetal life, and in embryoid bodies. Moreover, Risau and co-workers (Kappel et al., 2000; Rönicke et al., 1996) performed a functional analysis of the flk-1/VEGFR-2 promoter in vivo, thus demonstrating that flk-1 5′-flanking sequences confer endothelium-specific expression in transfected endothelial cells, and also identified by mutational analysis the binding sites for SCL/Tal-1, GATA, and Ets transcription factors as critical elements for the endothelium-specific flk-1/VEGFR-2 gene expression in transgenic mice, providing the first evidence that SCL/Tal-1, GATA, and Ets transcription factors act upstream of flk-1/VEGFR-2 in a combinatorial manner to determine embryonic blood vessel formation. SCL/Tal-1 activates VEGF expression in the endoderm and the mesoderm. In response, angioblasts upregulate the expression of surface flk-1/VEGFR-2 and migrate from the splanchnopleuric mesoderm and differentiate in response to paracrine fibroblast growth factor-2 (FGF-2) and VEGF signaling (Coultas et al., 2005). The loss of both vascular and hematopoietic cells in flk-1/VEGFR-2 mutants suggests that endothelial and hematopoietic developments are linked through a common progenitor, the hemangioblasts.

Genetic loss of Flt-1/VEGFR-1 leads to vessel dysmorphogenesis and overgrowth (Fong et al., 1995; Kearney et al., 2002). Fong et al. (1995)

showed that Flt-1/VEGFR-1 null mutant mice die in utero due to an overgrowth of endothelial cells and disorganization of blood vessels, suggesting that Flt-1/VEGFR-1 has a negative regulatory role in vascular development. However, mice expressing the VEGFR-1 extracellular and transmembrane domains, but lacking the tyrosine kinase domain, develop a normal vasculature (Hiratsuka et al., 1998).

Chappell et al. (2009) demonstrated that an analysis of developing vessels with perturbed Flt-1/VEGFR-1 function revealed misguided emerging sprouts and transgenic soluble Flt-1 (sFlt-1) rescued sprout guidance parameters. Studies using receptor-selective VEGF mutants demonstrated that VEGFR-1 cooperates with VEGFR-2 in inducing gene expression in endothelial cells in vitro although no unique expression pattern was observed in response to VEGFR-1 alone (Yang et al., 2002). The imbalance between the functions of VEGFR-1 and VEGFR-2 is responsible for the formation of hemangiomas (Junnin et al., 2008).

VEGFR-1 expression is induced by Notch signaling to reduce VEGF ligand availability, thus preventing tip cell outward migration. VEGFR-1 is predominantly expressed in stalk cells and is involved in guidance and limiting tip cell formation, and loss of VEGFR-1 increases sprouting and vascularization (Chappell et al., 2009). During both mouse and zebrafish angiogenesis ,VEGFR-3 is most strongly expressed in the leading tip cell and is downregulated by Notch signaling in the stalk cell (Shawber et al., 2007; Siekmann and Lawson, 2007).

The work from Kari Alitalo's (Fig. 3) laboratory in Helsinki was central to the characterization of VEGF-C/VEGFR-3, restricted largely to the lymphatic endothelium (Kukk et al., 1996; Ribatti, 2018). VEGFR-3 may play a role in disorders involving the lymphatic system and in angiogenesis

Fig. 3 A portrait of Kari Alitalo. (Reproduced from Ribatti D: Historical overview of lymphangiogenesis, *Curr Opin Immunol* 53:161–166, 2018.)

and may be of potential use in drug targeting, in vivo imaging of the lymphatic vessels, and in therapeutic lymphangiogenesis. VEGF-C binds to VEGFR-3, expressed on the lymphatic endothelium, and has been implicated in lymphangiogenesis. Overexpression of VEGF-C and VEGF-D in transgenic mice induces the formation of hyperplastic lymphatic vessels. Conversely, VEGFR-3 gene inactivation results in embryonic death due to abnormal remodeling of the primary vascular plexus (Dumont et al., 1998), and inhibition of VEGF-C and/or VEGF-D by overexpression of a soluble form of VEGFR-3 in the skin of transgenic mice leads to inhibition of lymphatic vessel growth (Jussila and Alitalo, 2002). Transgenic inactivation of both VEGF-C alleles results in prenatal death: endothelial cells commit to the lymphatic lineage but do not sprout from veins (Karkkainen et al., 2004).

Both angiopoietin-1 adn-2 (Ang-1 and Ang-2) mediate the balance between vascular quiescence and proliferation during the sprouting/migration phase of angiogenesis. Overexpression of Ang-2 in the embryos produces a phenotype similar to the loss of function of Ang-1 or Tie-2 (Maisonpierre et al., 1997). Transgenic overexpression of Ang-1 in the skin results in pronounced hypervascularization with the production of many compact stable vessels (Thurston et al., 1999; Suri et al., 1996). This contrasts with the effects of overexpression of VEGF, which also leads to hypervascularization but with the formation of large, leaky, simple endothelial tubes, unprotected by supporting cells (Drake and Little, 1995). If Ang-1 normally acts to stabilize vessels, then there has to be a mechanism to modulate its action to allow vascular remodeling and act to locally block Ang-1 action and destabilize vessels (Maisonpierre et al., 1997). If VEGF is also around, then local angiogenesis can occur. In the absence of VEGF, and in the presence of Ang-2, vessels destabilize and endothelial cell apoptosis occurs, resulting in vessel regression.

References

Ausprunk DH, Folkman J: Migration and proliferation of endothelial cells in preformed and newly formed blood vessels during tumor angiogenesis, *Microvasc Res* 14:53–65, 1977.

Breier G, Albrecht U, Sterrer S, et al.: Expression of vascular endothelial growth factor during embryonic angiogenesis and endothelial cell differentiation, *Development* 114:521–532, 1992.

Chappell JC, Taylor SM, Ferrara N, et al.: Local guidance of emerging vessel sprouts requires soluble flt-1, *Dev Cell* 17:377–386, 2009.

Cimpean AM, Ribatti D, Raica M: A brief history of angiogenesis assays, *Int J Dev Biol* 55:377–382, 2010.

Coultas L, Chawengsaksophak K, Rossant J: Endothelial cells and VEGF in vascular development, *Nature* 438:937–945, 2005.

Djonov VG, Galli AB, Burri PH: Intussusceptive arborization contributes to vascular tree formation in the chick chorio-allantoic membrane, *Anat Embryol* 202:347–357, 2000.

Drake CJ, Little CD: Exogenous vascular endothelial growth factor induces malformed and hyperfused vessels during embryonic neovascularization, *Proc Natl Acad Sci U S A* 92:7657–7761, 1995.

Dumont DJ, Jussila L, Taipale J, et al.: Cardiovascular failure in mouse embryos deficient in VEGF receptor-3, *Science* 282:946–949, 1998.

Folkman J, Haudenschild C: Angiogenesis in vitro, *Nature* 288:551–555, 1980.

Fong GH, Rossant J, Gertsenstein M, et al.: Role of the Flt-1 receptor tyrosine kinase in regulating the assembly of vascular endothelium, *Nature* 376:66–70, 1995.

Fruttiger M: Development of the retinal vasculature, *Angiogenesis* 10:77–88, 2007.

Grant DS, Lelkes PI, Fukuda K, et al.: Intracellular mechanisms involved in basement membrane induced blood vessel differentiation in vitro, *In Vitro Cell Dev Biol* 27:327–335, 1991.

Hiratsuka S, Minowa O, Kuno J, et al.: Flt-1 lacking the tyrosine kinase domain is sufficient for normal development and angiogenesis in mice, *Proc Natl Acad Sci U S A* 95:9349–9354, 1998.

Junnin M, Medici D, Park L, et al.: Suppressed NFAT-dependent VEGFR-1 expression and constitutive VEGFR-2 signaling in infantile hemangioma, *Nat Med* 14:1236–1246, 2008.

Jussila L, Alitalo K: Vascular growth factors and lymph angiogenesis, *Physiol Rev* 8:673–700, 2002.

Kabrun N, Buhring HJ, Choi K, et al.: Flk-1 expression defines a population of early embryonic hematopoietic precursors, *Development* 124:2039–2048, 1997.

Kappel A, Schlaeger TM, Flamme I, et al.: Role of SCL/Tal-1, GATA, and ets transcription factor binding sites for the regulation of flk-1 expression during murine vascular development, *Blood* 96:3078–3085, 2000.

Karkkainen MJ, Haiko P, Sainio K, et al.: Vascular endothelial growth factor C is required for sprouting of the first lymphatic vessels from embryonic veins, *Nat Immunol* 5:74–80, 2004.

Kearney JB, Ambler CA, Monaco KA, et al.: Vascular endothelial growth factor receptor Flt-1 negatively regulates developmental blood vessel formation by modulating endothelial cell division, *Blood* 99:2397–2407, 2002.

Kremer C, Breier G, Risau W, et al.: Up-regulation of flk-1/vascular endothelial growth factor receptor 2 by its ligand in a cerebral slice culture system, *Cancer Res* 57:3852–3859, 1997.

Kukk E, Lymboussaki A, Taira S, et al.: VEGF-C receptor binding and pattern of expression with VEGFR-3 suggests a role in lymphatic vascular development, *Development* 122:3829–3837, 1996.

Lawson ND, Weinstein BM: In vivo imaging of embryonic vascular development using transgenic zebrafish, *Dev Biol* 248:307–318, 2002.

Madri JA, Williams SK: Capillary endothelial cell cultures: phenotypic modulation by matrix components, *J Cell Biol* 97:153–165, 1983.

Maisonpierre PC, Suri C, Jines PF, et al.: Angiopoietin-2, a natural antagonist for Tie2 that disrupts in vivo angiogenesis, *Science* 277:55–60, 1997.

Millauer B, Wizigmann-Voos S, Schnürch H, et al.: High affinity VEGF binding and developmental expression suggest Flk-1 as a major regulator of vasculogenesis and angiogenesis, *Cell* 72:835–846, 1993.

Montesano R, Orci L, Vassalli P: In vitro rapid organization of endothelial cells into capillary-like networks is promoted by collagen matrices, *J Cell Biol* 97:1648–1652, 1983.

Muthukkaruppan V, Auerbach R: Angiogenesis in the mouse cornea, *Science* 206:1416–1418, 1979.

Paku S, Pawoletz N: First steps of tumor-related angiogenesis, *Lab Investig* 65:334–346, 1991.

Ribatti D: Genetic and epigenetic mechanisms in the early development of the vascular system, *J Anat* 208:139–152, 2006.

Ribatti D: Historical overview of lymphangiogenesis, *Curr Opin Immunol* 53:161–166, 2018.

Rönicke V, Risau W, Breier G: Characterization of the endothelium-specific murine vascular endothelial growth factor receptor-2 (Flk-1) promoter, *Circ Res* 79:277–285, 1996.

Shalaby F, Rossant J, Yamaguchi TP, et al.: Failure of blood-island formation and vasculogenesis in Flk-1-deficient mice, *Nature* 376:62–66, 1995.

Shawber CJ, Funahashi Y, Franciso E, et al.: Notch alters VEGF responsiveness in human and murine endothelial cells by direct regulation of VEGFR-3 expression, *J Clin Invest* 117:3369–3382, 2007.

Siekmann AF, Lawson ND: Notch signalling limits angiogenic cell behaviour in developing zebrafish arteries, *Nature* 445:781–784, 2007.

Suri C, Jones PF, Patan S, et al.: Requisite role of angiopoietin-1, a ligand for the Tie-2 receptor, during embryonic angiogenesis, *Cell* 87:1171–1180, 1996.

Thurston G, Suri C, Smith K, et al.: Leakage-resistant blood vessels in mice transgenically overexpressing angiopoietin-1, *Science* 286:2511–2514, 1999.

Vogeli KM, Jin SW, Martin GR, et al.: A common progenitor for haematopoietic and endothelial lineages in the zebrafish gastrula, *Nature* 443:337–339, 2006.

Yang S, Toy K, Ingle G, et al.: Vascular endothelial growth factor-induced genes in human umbilical vein endothelial cells, *Arterioscler Thromb Vasc Biol* 22:1797–1803, 2002.

CHAPTER 4

Nonsprouting angiogenesis: Intussusceptive microvascular growth

Morphologically, three forms of intussusceptive angiogenesis are known: intussusceptive microvascular growth (IMG), intussusceptive arborization (IAR), and intussusceptive branching remodeling (IBR). IMG renders rapid expansion of an already existing vascular network and its mechanism is most likely driven by blood flow. IAR remodels the capillary network into a hierarchical vascular tree pattern. IBR optimizes the vasculature by either changing the branching pattern of blood vessels or pruning the vascular network (Makanya et al., 2009; De Spiegelaere et al., 2012).

IMG, also known as nonsprouting angiogenesis, is an alternative or additional mechanism of capillary growth, whereby the vascular network expands by insertion of newly formed columns of interstitial tissue structures into the vascular lumen called tissue pillars or posts.

The first reports on IMG were published by Burri and co-workers who investigated the lung vasculature in postnatal rats using corrosion casting and scanning electron microscopy (Caduff et al., 1986; Burri and Tarek, 1990). Transluminal pillar formation is one particular method of expanding vessels (Djonov et al., 2000). According to Burri's group, this mechanism proceeds through four steps: (1) Protrusion of opposing capillary walls into the lumen and creation of a contact zone between facing endothelial cells (pillar perforation). The contacting endothelial cells can be prominent or planar, and the contacts can be symmetric or asymmetric. (2) Reorganization of their intercellular junctions and central perforation of the endothelial bilayer. (3) Formation of an interstitial pillar core by invading supporting cells (myofibroblasts, pericytes) and deposition of matrix, such as pillars ranging in diameters from 1 to 2.5 μm. (4) Enlargement in thicknesses of the pillars without additional qualitative alterations or merging with other pillars (Fig. 1) (Ribatti and Djonov, 2012). Hemodynamic conditions have a critical and rapid action on vascular adaptation and formation of intravascular

New Insights on the Development of the Vascular System
https://doi.org/10.1016/B978-0-323-90599-2.00004-0

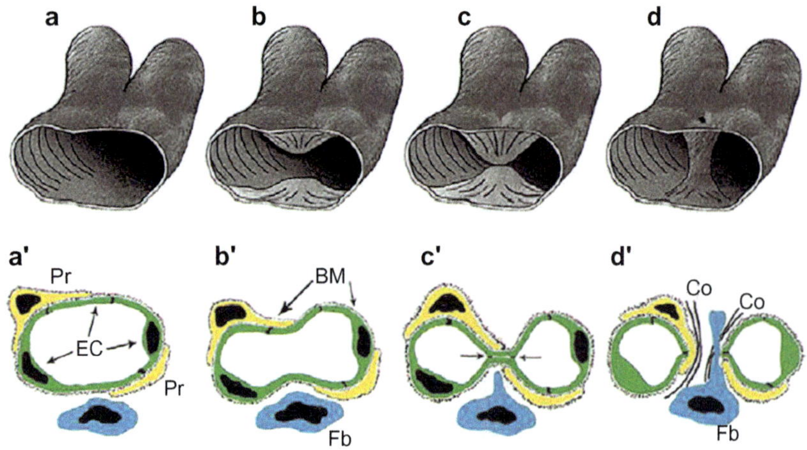

Fig. 1 3D (A–D) and 2D (A′–D′) schemes depicting the generation of transluminal pillars by intussusceptive angiogenesis. Simultaneous protrusion of opposing capillary walls into the vessel lumen (A, B; A′, B′) results in creation of an interendothelial contact zone (C; C′). In the subsequent step, the endothelial bilayer becomes perforated and the newly formed pillar core gets invaded by fibroblasts (Fb) and pericytes (Pr), which lay down collagen fibrils (Co in D′). (Reproduced from Ribatti D, Djonov V: Intussusceptive microvascular growth in tumors. *Cancer Lett* 316:126–131, 2012.)

pillars, as well as on vessel diameter (Djonov et al., 2002; Le Noble et al., 2004). The molecular mechanism of IMG is not fully understood, but shear stress plays a major role, as well as increasing blood flow rate. The role of hemodynamics in control of IMG has been demonstrated by clamping of one of the dichotomous branches of an artery in the developing chorioallantoic membrane (CAM) of the chick embryo (Djonov et al., 2002). Increase in blood flow and pressure in the cognate artery resulted in an almost immediate effect on branching morphology with pillars beginning to appear in 15–30 min of clamping (Djonov et al., 2002).

Baum et al. (2010) investigated the impact of vascular endothelial growth factor (VEGF) on IMG in the CAM vasculature. Light microscopy of the microvascular casts revealed the presence of hollows (>2lm in diameter) and minute holes (<2lm in diameter). The latter correspond to transluminal endothelial pillars, which are a morphological expression of intussusception (Djonov et al., 2000), whereas the hollows represent the impressions of strands of either mesenchymal cells or chorionic epithelial cells that traverse the capillary bed. Morphometric analysis revealed that the numerical density of the holes/pillars peak on days 8 and 12–14 of incubation, while the percentage area of the CAM that was occupied by epithelial strands corresponding to nonvascularized areas increases between days 8 and 10 of incubation. The transcription rate of VEGF-A mRNA peaked on days 8, 9, and 12 of incubation, whereas VEGF-A protein expression increased on days 8–9 and 11–12 of incubation to rapidly decrease thereafter as determined by immunoblotting. When the VEGF receptor (VEGFR) inhibitor PTTK787/ZK22584 was applied on the CAM on day 9 of incubation, the microvasculature topology on day 12 was similar to that on day 10. This study clearly demonstrated that the temporal course of intussusception corresponded to the expression of VEGF in the CAM microvasculature and that inhibition of VEGF signaling retarded intussusceptive-dependent capillary maturation.

Recombinant human erythropoietin (Epo) induces IMG in vivo (Crivellato et al., 2004). In response to Epo, three morphological structural changes in terms of expression of IMG have been identified: pillar formation by folding of the lateral vascular wall, fusion of pillars, and connection of intraluminal tissue folds with the opposite vascular wall. In some instances, a single endothelial cell connected the opposite walls of a preexisting vessel, like a transluminal bridge, for example, expression of the connection of intraluminal tissue folds with the opposite vascular wall. This process caused the splitting of the original vascular structure into two newly formed blood vessels. This response is peculiar of Epo, because it is abolished when an Epo-blocking antibody is coadministered with Epo (Crivellato et al., 2004).

IMG has been observed in capillary plexuses, small arteries, and veins. It has been described in human adenocarcinoma xenografts (Patan et al., 1996), mammary gland tumors (Djonov et al., 2001), B-cell non-Hodgkin's lymphomas (Crivellato et al., 2003), primary melanomas (Ribatti et al., 2005), and gliomas (Fig. 2) (Nico et al., 2010). Nico et al. (2010) correlated in human gliomas the extent of angiogenesis, evaluated as microvascular density, immunoreactivity of tumor cells to vascular VEGF, vessel diameter, and IMG to the tumor stage, and demonstrated a relationship in human glioma progression between angiogenesis, VEGF immunoreactivity of tumor cells, vessel diameter, and the number of connections of intraluminal tissue folds with the opposite vascular wall and expression of IMG, thus suggesting that IMG could be a mechanism of compensatory vascular growth occurring in human gliomas. High VEGF levels can induce endothelial cells that undergo IMG to either sprout or fuse (Drake and Little, 1999).

IMG generates vessels more rapidly with a less metabolic demand as compared to sprouting angiogenesis and is a putative strategy that tumors can use for rapid adaptation to milieu changes. In fact, no endothelial cell

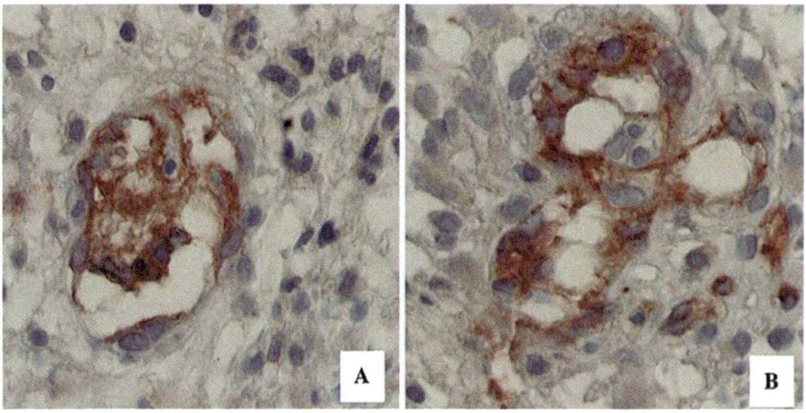

Fig. 2 Two examples of tumor vessels with a low and high number of connections of intraluminal tissue folds with the opposite vascular wall, i.e., expression of intussusceptive microvascular growth in a II grade malignant tumor specimen (A) when compared with a IV grade malignant tumor specimen (B). (Reproduced from Nico B, Crivellato E, Guidolin D, et al.: Intussusceptive microvascular growth in human glioma, *Clin Exp Med* 10:93–98, 2010.)

proliferation is required for IMG and endothelial cells only increase their volume and become thinner. Paku et al. (2011) elucidated the mechanism of pillar formation in experimental tumors. Using electron and confocal microscopy, they observed intraluminal nascent pillars that contain a collagen bundle covered by endothelial cells and proposed a new mechanism for the development of pillars consisting of four steps: (1) formation of intraluminal endothelial bridges; (2) on the abluminal side of the endothelial cells that form the bridge, the basement membrane is locally disrupted by proteolytic activity; (3) an endothelial cell from the bridge adheres to a nearly collagen bundle, which is transferred through the lumen, reaches the other side of the lumen, and is transferred into the connective tissue on the other side of the vessel; and (4) further pillar maturation occurs through immigration of fibroblasts/myofibroblasts and pericytes into the pillars and subsequent extracellular matrix protein (collagen and fibrin) deposition by these cells.

References

Baum O, Suter F, Gerber B, et al.: VEGF-A promotes intussusceptive angiogenesis in the developing chicken chorioallantoic membrane, *Microcirculation* 17:447–457, 2010.

Burri PH, Tarek MR: A novel mechanism of capillary growth in the rat pulmonary microcirculation, *Anat Rec* 228:35–45, 1990.

Caduff JH, Fischer LC, Burri PH: Scanning electron microscopic study of the developing microvasculature in the postnatal rat lung, *Anat Rec* 216:154–164, 1986.

Crivellato E, Nico B, Vacca A, et al.: B-cell non-Hodgkin's lymphomas express heterogeneous patterns of neovascularization, *Haematologica* 88:671–678, 2003.

Crivellato E, Nico B, Vacca A, et al.: Recombinant human erythropoietin induces intussusceptive microvascular growth in vivo, *Leukemia* 18:331–336, 2004.

De Spiegelaere WC, Casteleyn C, Van den Broek W, et al.: Intussusceptive angiogenesis: a biologically relevant form of angiogenesis, *J Vasc Res* 49:390–404, 2012.

Djonov V, Schmid M, Tschanz SA, et al.: Intussusceptive angiogenesis: its role in embryonic vascular network formation, *Circ Res* 86:286–292, 2000.

Djonov V, Andres AC, Ziemiecki A: Vascular remodeling during the normal and malignant life cycle of the mammary gland, *Microsc Res Tech* 52:182–189, 2001.

Djonov VG, Kurz H, Burri PH: Optimality in the developing vascular system: bifurcation remodeling by means of intussusception as an efficient adaptation mechanism, *Dev Dyn* 224:391–402, 2002.

Drake CJ, Little CD: VEGF and vascular fusion: implications for normal and pathological vessels, *J Histochem Cytochem* 74:1351–1355, 1999.

Le Noble F, Moyon D, Pardanaud L, et al.: Flow regulates arterial-venous differentiation in the chick embryo yolk sac, *Development* 131:361–375, 2004.

Makanya AN, Hlushchuk R, Djonov V: Intussuseptive angiogenesis and its role in vascular morphogenesis, patterning and remodeling, *Angiogenesis* 12:113–123, 2009.

Nico B, Crivellato E, Guidolin D, et al.: Intussusceptive microvascular growth in human glioma, *Clin Exp Med* 10:93–98, 2010.

Paku S, Deszo K, Bugyik E, et al.: A new mechanism for pillar formation during tumor-induced intussusceptive angiogenesis. Inverse sprouting, *Am J Pathol* 179:1573–1585, 2011.

Patan S, Munn LL, Jain RK: Intussusceptive microvascular growth in a human colon adenocarcinoma xenograft: a novel mechanism of tumor angiogenesis, *Microvasc Res* 53:33–52, 1996.

Ribatti D, Djonov V: Intussusceptive microvascular growth in tumors, *Cancer Lett* 316:126–131, 2012.

Ribatti D, Nico B, Floris C, et al.: Microvascular density, vascular endothelial growth factor immunoreactivity in tumor cells, vessel diameter and intussusceptive microvascular growth in primary melanoma, *Oncol Rep* 14:81–84, 2005.

CHAPTER 5

Endothelial tip, stalk, and phalanx cells

Initiation of sprouting requires the specification of endothelial cells into tip and stalk cells bearing different morphologies and functional properties. Endothelial tip cells primarily migrate but proliferate only minimally, in contrast to endothelial stalk cells, which do proliferate. In 2003, the concept of the "tip" and "stalk" cell phenotype was described for emerging sprouts (Gerhardt et al., 2003), even though filopodia-studded cells at the front were already described before this work by Kurz et al. (1996) and, later, by Ruhrberg et al. (2002). According to this "new" model, two principal cells are involved in sprouting angiogenesis, namely, "tip cell" and "stalk cell" (Fig. 1). Tip cells express high levels of delta-like ligand-4 (Dll-4) compared with those of stalk cells, which have higher levels of Notch signaling activity. Dll-4 is expressed in tip cells, which binds to Notch receptors on stalk cells, thus preventing their transformation into tip cells. Notch signaling induces the transition of tip cells to stalk cells, whereas inhibition of Notch signaling results in a dramatic increase in sprouting, branching, and extension of filopodia by tip cells.

Tip cells were identified by Christer Betzholtz and Hoger Gerhard of the Karolinska Institute in Stockholm (Gerhardt et al., 2003). The tip cell is migratory and polarized with a minimal proliferation rate, whereas the stalk cell proliferates during sprout extension and forms a nascent vascular lumen cell by maintaining its position and connection to the parent vasculature.

The phenotypic specialization of endothelial cells into tip or stalk cells is extremely transient and reversible, depending on the balance between proangiogenic factors, such as vascular endothelial growth factor (VEGF) and Jagged-1 (JAG-1), and suppressors of endothelial cell proliferation, such as Dll-4-Notch activity (Eilken and Adams, 2010; Geudens and Gerhardt, 2011; Wacker and Gerhardt, 2011; Tung et al., 2012). JAG-1 is expressed primarily by stalk cells that are involved in activating Notch-1 and maintains differential Notch activity by antagonizing Dll-4 that signals back to tip cells (Potente et al., 2011).

New Insights on the Development of the Vascular System
https://doi.org/10.1016/B978-0-323-90599-2.00005-2

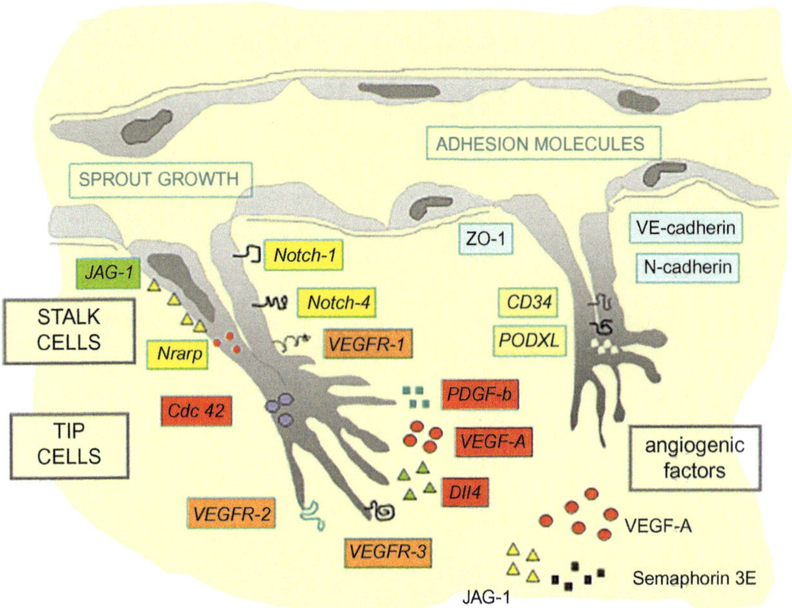

Fig. 1 Functional specialization of endothelial cells during the sprouting process. VEGF-A induces the formation and extension of filopodia as well as the expression of Dll-4 protein in tip cells. Tip cells express high levels of Dll-4, PDGF-B, UNC5b, VEGFR-2, and VEGFR-3/Flt-4 and have low levels of Notch signaling activity. During both mouse and zebrafish angiogeneses, VEGFR-3 is most strongly expressed in the leading tip cell and is downregulated by Notch signaling in the stalk cell. Stalk cells produce fewer filopodia, are more proliferative, form tubes and branches, and also form a vascular lumen. They also establish junctions with neighboring cells and synthesize basement membrane components. Stalk cells have high levels of Notch signaling activity and elevated expression of JAG-1. Nrarp is downstream of Notch and counteracts Notch signaling and is expressed in stalk cells at the branch points. Adherens junction formation is associated with the inhibition of endothelial cell migration in monolayers. This process is mediated by VE-cadherin. Another cell adhesion molecule expressed in endothelial cells is N-cadherin. VE-cadherin is strictly required for the polarization of endothelial cells. Endothelial cell polarization starts with the delivery of de-adhesive apical glycoproteins to the cell-cell contact via exocytosis. De-adhesive molecules include CD34 sialomucins, such as CD34 and PODXL. (Reproduced from Ribatti D, Crivellato E: Sprouting angiogenesis: a reappraisal, *Dev Biol* 372:157–165, 2012.)

VEGF-induced Notch signaling regulates the acquisition of the tip cell identity and tip-stalk conversion. At the angiogenic front where the VEGF level is high, some endothelial cells acquire tip cell identity through a competition mechanism (Ubezio et al., 2016). When an endothelial cell is activated by VEGF, it expresses Dll-4, leading to activation of Notch signaling in neighboring cells, which, in turn, suppresses VEGFR-2 expression.

Tip cells express high levels of Dll-4, platelet-derived growth factor-B (PDGF-B), UNC5 homolog b (UNC5b), VEGF receptor-2 (VEGFR-2), and VEGFR-3/Flt-4 and have low levels of Notch signaling activity (Gerhardt et al., 2003; Claxton and Fruttiger, 2004; Lu et al., 2004; Siekmann and Lawson, 2007a,b). The zebrafish tip cell is a highly branched structure (Siekmann and Lawson, 2007a,b), whereas in other mouse angiogenesis models, such as the brain and the retina (Gerhardt et al., 2003, 2004; Dorrell and Friedlander, 2006), the tip cell extends numerous filopodia, which serve to guide the new blood vessel in a certain direction toward an angiogenic stimulus, proliferates minimally, and adopts a highly branched shape while moving (Isogai et al., 2003). Bentley et al. (2009) developed a new physics-based model tightly coupled to in vivo data and proposed a novel filopodia adhesion-driven migration mechanism, demonstrating that cell-cell junction size is a key factor in establishing a stable tip/stalk pattern.

Stalk cells produce fewer filopodia, are more proliferative, form tubes and branches, and also form a vascular lumen (Thurston and Kitajewski, 2008). They also establish junctions with neighboring cells and synthesize basement membrane components (Phng and Gerhardt, 2009). Only a regulated balance between both processes establishes adequately shaped nascent sprouts (Eilken and Adams, 2010; Geudens and Gerhardt, 2011; Wacker and Gerhardt, 2011; Tung et al., 2012).

During the transition from active sprouting to quiescence endothelial cells, tip cells adopt a "phalanx" phenotype, resembling the phalanx formation of ancient Greek soldiers; these cells are lumenized, nonproliferating, and immobile, which promotes vessel integrity and stabilizes the vasculature through increased cell adhesion and dampened response to VEGF (Bautch, 2009; Mazzone et al., 2009). Phalanx cells are closely connected by tight junctions and adherens junctions and form a barrier between blood and surrounding tissues (Potente et al., 2011).

Tip cell migration depends on a gradient of VEGF, whereas stalk cell proliferation is regulated by VEGF concentration (Gerhardt et al., 2003). The leading tip cell responds to a VEGF gradient by migrating outward from the parent vessel up the gradient (Ruhrberg et al., 2002). VEGF induces the formation and extension of filopodia as well as the expression of Dll-4 protein in tip cells, and filopodia engages with those of a nearby tip cell to form a "bridge" and in the formation of a new vessel (Bentley et al., 2009). As the vessel elongates, the stalk cell proliferates in response to VEGF, creates a lumen, synthesizes a basement membrane, associates with pericytes, and increases the mass and surface of the growing vessel.

VEGF stimulates tip cell induction and filopodia formation via VEGFR-2, which is abundant on filopodia, whereas VEGFR-2 blockade is associated with sprouting defects (Phng and Gerhardt, 2009). Moreover, activation of Cdc 42 by VEGF triggers filopodia formation (De Smet et al., 2009). VEGFR-1 expression is induced by Notch signaling to reduce VEGF ligand availability, thus preventing tip cell outward migration. VEGFR-1 is predominantly expressed in stalk cells and is involved in guidance and in limiting tip cell formation, and loss of VEGFR-1 increases sprouting and vascularization (Chappell et al., 2009; Chappell and Bautch, 2010). During both mouse and zebrafish angiogeneses, VEGFR-3 is most strongly expressed in the leading tip cell and is downregulated by Notch signaling in the stalk cell (Shawber et al., 2007; Siekmann and Lawson, 2007a,b).

Endothelial cell migration is regulated by additional members of Rho GTPase related to RhoA and RAC1 (Nohata et al., 2016; Zahra et al., 2019). Moreover, RhoJ is important for the integration of VEGF and SEMA3A signals involved in the control of endothelial cell migration (Fukushima et al., 2020). Coordination of endothelial cell movements is associated with lamellipodia-like protrusions emerging from endothelial cell junctions, including junction-associated lamellipodia (JAIL) and junction-based lamellipodia (Paatero et al., 2018; Cao et al., 2007). Moreover, the dedicator of cytokinesis (DOCK) protein family is involved in vascular development through its regulation of Cdc 42 and RAC1 activity (Benson and Southgate, 2021).

References

Bautch VL: Endothelial cells form a phalanx to block metastasis, *Cell* 136:810–812, 2009.

Benson CE, Southgate L: The DOCK protein family in vascular development and disease, *Angiogenesis* 24:417–433, 2021.

Bentley K, Mariggi G, Gerhardt H, et al.: Tipping the balance: robustness of tip selection, migration and fusion in angiogenesis, *PLoS Comput Biol* 5, 2009, e1000549.

Cao Y, Sonveaux P, Liu S, et al.: Systemic overexpression of angiopoietin-2 promotes tumor microvessel regression and inhibits angiogenesis and tumor growth, *Cancer Res* 67:3835–3844, 2007.

Chappell JC, Bautch VL: Vascular development: genetic mechanisms and links to vascular diseases, *Curr Top Dev Biol* 90:43–72, 2010.

Chappell JC, Taylor SM, Ferrara N, et al.: Local guidance of emerging vessel sprouts requires soluble flt-1, *Dev Cell* 17:377–386, 2009.

Claxton S, Fruttiger M: Periodic delta-like 4 expression in developing retinal arteries, *Gene Expr Patterns* 5:123–127, 2004.

De Smet F, Segura I, De Bock K, et al.: Mechanisms of vessel branching: filopodia of endothelial tip cells lead the way, *Arterioscler Thromb Vasc Biol* 29:639–649, 2009.

Dorrell MI, Friedlander M: Mechanisms of endothelial cell guidance and vascular patterning in the developing mouse retina, *Prog Retin Eye Res* 25:277–295, 2006.

Eilken HM, Adams RH: Dynamics of endothelial cell behavior in sprouting angiogenesis, *Curr Opin Cell Biol* 22:617–625, 2010.

Fukushima Y, Nishiyama K, Kataoka H, et al.: RhoJ integrates attractive and repulsive cues in directional migration of endothelial cells, *EMBO J* 39, 2020, e102930.

Gerhardt H, Golding M, Fruttiger M, et al.: VEGF guides angiogenic sprouting utilizing endothelial tip cell filopodia, *J Cell Biol* 161:1163–1177, 2003.

Gerhardt H, Ruhrberg C, Abramsson A, et al.: Neuropilin-1 is required for endothelial tip cell guidance in the developing central nervous system, *Dev Dyn* 231:503–509, 2004.

Geudens I, Gerhardt H: Coordinating cell behavior during blood vessel formation, *Development* 138:4569–4583, 2011.

Isogai S, Lawson ND, Torrealday S, et al.: Angiogenic network formation in the developing vertebrate trunk, *Development* 130:5281–5290, 2003.

Kurz H, Gätmer T, Eggli PS, et al.: First blood vessels in the avian neural tube are formed by a combination of dorsal angioblasts immigration and ventral sprouting of endothelial cells, *Dev Biol* 173:133–147, 1996.

Lu X, Le Noble F, Yuan L, et al.: The netrin receptor UNC5B mediates guidance events controlling morphogenesis of the vascular system, *Nature* 432:179–186, 2004.

Mazzone M, Dettori D, de Oliveira L, et al.: Heterozygous deficiency of PHD2 restores tumor oxygenation and inhibits metastasis via endothelial normalization, *Cell* 136:839–851, 2009.

Nohata N, Uchida Y, Stratman AN, et al.: Temporal-specific roles of Rac1 during vascular development and retinal angiogenesis, *Dev Biol* 411:183–194, 2016.

Paatero I, Sauteur L, Lee M, et al.: Junction-based lamellipodia drive endothelial cell rearrangements in vivo via a VE-cadherin-F-actin based oscillatory cell-cell interaction, *Nat Commun* 9:3545, 2018.

Phng LK, Gerhardt H: Angiogenesis: a team effort coordinated by notch, *Dev Cell* 16:196–208, 2009.

Potente M, Gerhardt H, Carmeliet P: Basic and therapeutic aspects of angiogenesis, *Cell* 146:873–887, 2011.

Ruhrberg C, Gerhardt H, Golding M, et al.: Spatially restricted patterning cues provided by heparin-binding VEGF-A control blood vessel branching morphogenesis, *Cancer Dev* 16:2684–2698, 2002.

Shawber CJ, Funahashi Y, Franciso E, et al.: Notch alters VEGF responsiveness in human and murine endothelial cells by direct regulation of VEGFR-3 expression, *J Clin Invest* 117:3369–3382, 2007.

Siekmann AF, Lawson ND: Notch signalling and the regulation of angiogenesis, *Cell Adhes Migr* 1:104–106, 2007a.

Siekmann AF, Lawson ND: Notch signalling limits angiogenic cell behaviour in developing zebrafish arteries, *Nature* 445:781–784, 2007b.

Thurston G, Kitajewski J: VEGF and delta-notch: interacting signaling pathways in tumor angiogenesis, *Br J Cancer* 99:1204–1209, 2008.

Tung JJ, Tattersall IW, Kitajewski J: Tips, stalks, tubes: notch-mediated cell fate determination and mechanisms of tubulogenesis during angiogenesis, *Cold Spring Harb Perspect Med* 4, 2012, a006601.

Ubezio B, Bianco RA, Geudens I, et al.: Synchronization of endothelial dll4-notch dynamics switch blood vessels from branching to expansion, *elife* 5, 2016, e12167.

Wacker A, Gerhardt H: Endothelial development taking shape, *Curr Opin Cell Biol* 23:1–10, 2011.

Zahra FT, Sajib MS, Ichiyama Y, et al.: Endothelial RhoA GTPase is essential for in vitro endothelial functions but dispensable for physiological in vivo angiogenesis, *Sci Rep* 9:11666, 2019.

CHAPTER 6

Role of neuropilins

Neuropilins (NRPs), a small family of transmembrane proteins, were discovered as neuronal receptors for the class III semaphorin guidance cue. They are also expressed by tumor and endothelial cells and regulate angiogenesis. NRPs bind certain members of the vascular endothelial growth factor (VEGF) family: NRP-1 and NRP-2 bind VEGF-165 and placental growth factor (PlGF); NRP-1 also binds VEGF-B; and NRP-2 also binds VEGF-145 and VEGF-C (Karkkainen et al., 2001; Neufeld et al., 2002). Both NRP-1 and NRP-2 have about 50% sequence homology and similar domain structures (Neufeld et al., 2002).

The VEGF-binding ability of NRPs suggests that they may function as regulators of vasculogenesis. Mice lacking a functional NRP-1 gene display vascular abnormalities (Kawasaki et al., 1999), including impaired neural vascularization and defects in the development of large arteries, such as the branchial arch arteries. In addition, the development of the heart was strongly impaired, and heart failure was responsible for premature death. The vasculature of mice lacking a functional NRP-2 gene develops almost normally, except for minor defects in the lymphatic vessels (Giger et al., 2000; Yuan et al., 2002). NRP-2 does play an important role in vasculogenesis because mice lacking it display vascular abnormalities, which are much more severe than those seen in mice lacking a functional NRP-1 gene (Takashima et al., 2002). These phenotypes suggest a continued requirement for VEGF signaling to promote vascularization of developing organs, such as the brain and the heart, as well as an expansion and growth of major blood vessels. Overexpression of NRP-1 in transgenic embryos led to excess and dilation of blood vessels and heart malformation (Kitsukawa et al., 1995), consistent with NRP-1 acting as a coactivator of the VEGF receptor.

References

Giger RJ, Clouitier JF, Sahay A, et al.: Neuropilin-2 is required in vivo for selective axon guidance responses to secreted semaphorins, *Neuron* 25:29–41, 2000.

New Insights on the Development of the Vascular System
https://doi.org/10.1016/B978-0-323-90599-2.00006-4

Karkkainen MJ, Saaristo A, Jussila L, et al.: A model for gene therapy of human hereditary lymphedema, *Proc Natl Acad Sci U S A* 98:12677–12682, 2001.
Kawasaki T, Kitsukawa T, Bekku Y, et al.: A requirement for neuropilin-1 in embryonic vessel formation, *Development* 126:4895–4902, 1999.
Kitsukawa T, Shimono A, Kawakami A, et al.: Overexpression of a membrane protein, neuropilin, in chimeric mice causes anomalies in the cardiovascular system, nervous system and limbs, *Development* 121:4309–4318, 1995.
Neufeld G, Cohen T, Shraga N, et al.: The neuropilins: multifunctional semaphorin and VEGF receptors that modulate axon guidance and angiogenesis, *Trends Cardiovasc Med* 12:13–19, 2002.
Takashima S, Kitakaze M, Asakura M, et al.: Targeting of both mouse neuropilin-1 and neuropilin-2 genes severely impairs developmental yolk sac and embryonic angiogenesis, *Proc Natl Acad Sci U S A* 99:3657–3662, 2002.
Yuan L, Moyon D, Pardanaud L, et al.: Abnormal lymphatic vessel development in neuropilin 2 mutant mice, *Development* 129:4797–4806, 2002.

CHAPTER 7

Role of notch and notch ligands

Notch signaling is a highly conserved pathway, which was initially discovered in Drosophila development, and plays a crucial role in development and malignant transformation (Baron et al., 2002). Notch signaling is activated by interactions between Notch receptors, large highly conserved transmembrane proteins, and their ligands, namely, Delta, Serrate (Jagged in mammals), and Lag-2 (DSL). Notch-1 and -4 are specific to mouse arterial endothelial cells, whereas Notch-1a/b and -3 are expressed in zebrafish arteries. Mice possess five Notch ligands, Delta-like (Dll)-1, Dll-3, Dll-4, Jagged-1, and -2 (JAG-1 and JAG-2). There are four Notch receptors (Notch-1–4) and five ligands (Jagged-1, -2, Delta-1, -3, and -4) (Iso et al., 2003). Notch receptors are large transmembrane proteins, which, when activated by their ligands expressed on adjacent cells, regulate cell fate in multiple lineages.

After Notch signaling is activated by Dll or Jagged, the Notch intracellular domain (NICD) is cleaved and released and then enters the nucleus and binds with the CBF1/RBP-JKappa, Suppressor of hairless, Lag-1 (CSL) family of transcription factors. The Notch signaling pathway is essential for vascular development. Notch signaling-deficient embryos present a poorly differentiated dorsal aorta and a posterior cardinal vein associated with arteriovenous malformations, including the fusion of arteries and veins without an intervening capillary bed (Lawson et al., 2001). All the receptors and ligands are expressed in at least one vascular compartment, e.g., the arteries, veins, capillaries, muscle cells, or pericytes. Notch signaling is required for remodeling the primary vascular plexus into the hierarchy of mature vascular beds and for maintaining arterial fate (Alva and Iruela-Arispe, 2004). The necessity of Dll-4/Notch-1 signaling in the endothelium has been well established as the loss of a single copy of Dll-4, and deletion of Notch-1 causes vascular defects and embryonic lethality (Gale et al., 2004; Limbourg et al., 2005). In Shh-deficient embryos, Dll-4 expression is reduced and a defective dorsal aorta formation in Shh-deficient embryos can be rescued by increasing vascular endothelial growth factor (VEGF) signaling (Coultas et al., 2010). Notch-1 and Notch-4 and three Notch ligands, namely,

New Insights on the Development of the Vascular System
https://doi.org/10.1016/B978-0-323-90599-2.00007-6

JAG-1, Dll-1, and Dll-4, are expressed in endothelial cells for the induction of arterial cell fate and for the selection of endothelial tip and stalk cells during sprouting angiogenesis (Kume, 2009). Notch is also activated by arterial levels of laminar fluid shear stress, which contributes to suppression of endothelial cell proliferation (Fang et al., 2017). A loss of Notch signaling induces sprouting, whereas its activation reduces it. Notch-1-deficient endothelial cells adopt tip cell characteristics (Hellström et al., 2007a,b), while Notch signaling activity is greater in stalk cells, in which activation of Notch by Dll-4 leads to a downregulation of VEGF receptor-2 and -3 (VEGFR-2 and -3) in these cells (Hofmann and Iruela-Arispe, 2007). Cells dynamically compete for the tip position utilizing differential VEGFR levels, as cells with higher VEGFR signaling produce more Dll-4 and therefore inhibit their neighboring cells. In this context, a fine-tuned feedback loop between VEGF and Notch/Dll-4 signaling pathways is established and a crosstalk between these pathways is essential for proper patterning of the vasculature (Holderfield and Hughes, 2008).

Blockade of Notch leads to widespread Flt-4 expression, increases filopodia and sprouting, and promotes tip cell activity (Thurston et al., 2007; Siekmann and Lawson, 2007a). Tip cells with low Notch activity have high VEGFR-2 and low VEGFR-1 expressions, which result in higher levels of Dll-4 expression and, hence, through a higher production of Dll-4 than in neighboring cells, in an increased ability to suppress their neighboring cells from becoming tip cells (Jakobbson et al., 2010). Endothelial cells with higher levels of VEGF increase Dll-4 expression, which further increases the cell sensitivity to VEGF, and this cell becomes the tip cell selected for outward migration for the parent vessel (Hellström et al., 2007a,b; Lobov et al., 2007; Sainson et al., 2005; Siekmann and Lawson, 2007b; Tammela et al., 2008). Stalk cells have high levels of Notch signaling activity and elevated expression of JAG-1 (Benedito et al., 2009). Stalk cell JAG-1 antagonizes Dll-4 activity, reducing the induction of Notch signaling in the adjacent tip cell, which therefore maintains its responsiveness to VEGF stimulation and migrates outward to establish a new branch (Benedito et al., 2009). However, JAG-1 poorly activates Notch-1, as modification of Notch by FRINGE glycosyltransferase favors activation by Dll-4 (Eilken and Adams, 2010). Notch-regulated ankyrin repeat protein (Nrarp) is downstream of Notch and counteracts Notch signaling and is expressed in stalk cells at the branch points (Phng et al., 2009). Silencing of Nrarp is responsible for reduction in vessel density, due to a lowered endothelial cell proliferation as a consequence of an upregulation of Notch and VEGFR-1, of poorly lumenized vessels, remodeling of endothelial junctions, and vessel regression. By inducing a Wnt signal in stalk cells, Nrarp is responsible for vascular stabilization and lowering the Wnt signal induces vessel regression (Phng et al., 2009).

References

Alva JA, Iruela-Arispe ML: Notch signaling in vascular morphogenesis, *Curr Opin Hematol* 4:278–283, 2004.

Baron M, Aslavm H, Flasza M, et al.: Multiple levels of notch signal regulation (review), *Mol Membr Biol* 19:27–38, 2002.

Benedito R, Roca C, Sorensen I, et al.: The notch ligand Dll4 and jagged 1 have opposing effects on angiogenesis, *Cell* 137:1124–1135, 2009.

Coultas L, Nieuwwnhuis E, Anderson GA, et al.: Hedgehog regulates distinct vascular patterning events through VEGF-dependent and -independent mechanisms, *Blood* 116:653–660, 2010.

Eilken HM, Adams RH: Dynamics of endothelial cell behavior in sprouting angiogenesis, *Curr Opin Cell Biol* 22:617–625, 2010.

Fang JS, Coon BG, Gillis N, et al.: Shear-induced notch-cx-37-p27 axis arrests endothelial cell cycle to enable arterial specification, *Nat Commun* 8:2149, 2017.

Gale NW, Dominguez MG, Noguera I, et al.: Haploinsufficiency of delta-like-4 ligand results in embryonic lethality due to major defects in arterial and vascular development, *Proc Natl Acad Sci U S A* 101:15949–15954, 2004.

Hellström M, Phng LK, Gerhardt H:VEGF and notch signaling: the yin and yang of angiogenic sprouting, *Cell Adhes Migr* 1:133–136, 2007a.

Hellström M, Phng LK, Hofmannn JJ, et al.: Dll4 signalling through notch1 regulates formation of tip cells during angiogenesis, *Nature* 445:776–780, 2007b.

Hofmann JJ, Iruela-Arispe ML: Notch signaling in blood vessels: who is talking to whom about what? *Circ Res* 100:1556–1568, 2007.

Holderfield MT, Hughes CW: Crosstalk between vascular endothelial growth factor, notch, and transforming growth factor-β in vascular morphogenesis, *Circ Res* 102:637–652, 2008.

Iso T, Hamamori Y, Kedes L: Notch signaling in vascular development, *Arterioscler Thromb Vasc Biol* 23:543–553, 2003.

Jakobbson L, Franco CA, Bentley K, et al.: Endothelial cells dynamically compete for the tip cell position during angiogenic sprouting, *Nat Cell Biol* 12:943–953, 2010.

Kume T: Novel insights into the differential functions of notch ligands in vascular formation, *J Angiogenes Res* 1:8, 2009.

Lawson ND, Scheer N, Fham VM, et al.: Notch signalling is required for arterial-venous differenetiation during embryonic vascular development, *Development* 128:3675–3683, 2001.

Limbourg FP, Yakeshita K, Radtke F, et al.: Essential role of endothelial Notch1 in angiogenesis, *Circulation* 111:1826–1832, 2005.

Lobov IB, Renard RA, Papadopoulos N, et al.: Delta-like ligand 4 (Dll4) is induced by VEGF as negative regulator of angiogenic sprouting, *Proc Natl Acad Sci U S A* 104:3219–3224, 2007.

Phng LK, Potente M, Leslie JD, et al.: 2Nrarp coordinates endothelial notch and wnt signaling to control vessel density in angiogenesis, *Dev Cell* 16:70–82, 2009.

Sainson RC, Aoto J, Nakatsu MN, et al.: Cell-autonomous notch signaling regulates endothelial cell branching and proliferation during vascular tubulogenesis, *FASEB J* 19:1027–1029, 2005.

Siekmann AF, Lawson ND: Notch signalling and the regulation of angiogenesis, *Cell Adhes Migr* 1:104–106, 2007a.

Siekmann AF, Lawson ND: Notch signalling limits angiogenic cell behaviour in developing zebrafish arteries, *Nature* 445:781–784, 2007b.

Tammela T, Zarkada G, Wallgard E, et al.: Blocking VEGFR-3 suppresses angiogenic sprouting and vascular newtwork formation, *Nature* 454:656–660, 2008.

Thurston G, Noguera-Troise I, Yancopoulos GD: The delta paradox: Dll4 blockade leads to more tumour vessels but less tumour growth, *Nat Rev Cancer* 7:327–331, 2007.

CHAPTER 8

Semaphorins and netrin-UNC5B signaling

Semaphorins (SEMAs) are a family of cell surface-attached, secreted, and membrane-bound guidance cues involved in both attractive and repulsive activities in the vascular and nervous system formation, which interact with multimeric receptor complexes formed by neuropilins (NRPs) alone or by NRP/plexin family proteins (Carmeliet and Tessier-Lavigne, 2005). Membrane-associated SEMAs bind to plexins, whereas secreted SEMAs bind to NRPs.

More than 20 SEMAs have been identified and categorized into 8 classes. SEMAs from classes 1 and 2 and SEMA5C are found only in the invertebrates. SEMAs that belong to classes 3–7 are found only in the vertebrates, and SEMAs of class 8 are found in viruses. Membrane-associated SEMAs bind directly to plexins, whereas SEMA3A–F require NRPs, which act as coreceptors for plexin signaling. Plexins are single-pass transmembrane receptors and are subdivided into four groups, namely, A, B, C, and D.

Angiostatic SEMAs include SEMA3A, SEMA3B, SEMA3F, interacting with NRP receptors, and SEMA3E and SEMA4D, interacting with plexin D1 receptors. SEMA3C, SEMA4A, SEMA4D, SEMA6D, and SEMA7A possess proangiogenic properties, interacting with plexins A and B (Neufeld and Kessler, 2008). SEMA3A is expressed by endothelial cells of developing vessels and inhibits endothelial migration by interfering with integrin function (Bates et al., 2003; Serini et al., 2003). SEMA3A antagonizes the effects of vascular endothelial growth factor (VEGF) in an in vitro angiogenesis assay and in vivo (Miao et al., 1999; Bates et al., 2003). SEMA3F can inhibit fibroblast growth factor-2 (FGF-2) and VEGF-induced angiogenesis in vivo (Kessler et al., 2004). Binding of SEMA3A to NRP-1 or that of SEMA3F to NRP-2 induces repulsion of endothelial filopodia and retraction and inhibits adhesion and migration (Neufeld and Kessler, 2008). Both SEMA3A and SEMA3F inhibit vascular remodeling during embryonic development through an effect on integrin-mediated cell adhesion (Serini et al., 2003). The remodeling of both cephalic plexus and dorsal longitudinal

New Insights on the Development of the Vascular System
https://doi.org/10.1016/B978-0-323-90599-2.00008-8

anastomotic vessels into a mature organized vascular tree is defective in SEMA3A knockout embryos (Serini et al., 2003).

SEMA4D is angiogenic in vitro and in vivo and this effect is mediated by a high-affinity receptor, plexin B1, and the biological effects elicited by plexin B1 require coupling and activation of the Met tyrosine kinase (Conrotto et al., 2005). SEMA4D induces endothelial cell migration and tubulogenesis in vitro and stimulates blood vessel formation in vivo (Basile et al., 2004). Loss of plexin D1 expressed by endothelial cells is responsible of an aberrant sprouting into SEMA3E-expressing tissues in zebrafish embryos (Adams and Eichmann, 2010). In the mouse retina, ααSEMA3E activates plexin D1 on tip cells to fine-tune the balance on tip and stalk cells necessary for even-growing vascular fronts by coordinating VEGF's activity in a negative feedback (Kim et al., 2011). SEMA6D is required for cardiac looping and ventricular ballooning (Toyofuku et al., 2004). SEMA3E produced by the lateral plate mesoderm is required for dorsal aorta patterning and for generating the avascular zones that are located laterally to the dorsal aortae and along the midline (Meadows et al., 2013). SEMA4A-deficient mice demonstrated increased blood vessel formation in response to the VEGF matrigel plug assay (Toyofuku et al., 2007).

Netrins are laminin-related and secreted bifunctional guidance cues, which also bind to extracellular matrix components: attraction and repulsion are mediated by binding to the deleted in colorectal cancer (DCC) and UNC5 family receptors, respectively (Dickson and Keleman, 2002). Netrins regulate axon guidance and act as bifunctional guidance cues: they can attract some axons, while repelling others. Lu et al. (2004) demonstrated that UNC5B is selectively expressed in the vascular system by the arteries, a subset of capillaries, and endothelial tip cells. UNC5B inactivation results in enhanced sprouting, whereas netrin-1 prompts filopodia retraction of endothelial cells (Adams and Eichmann, 2010). These results suggest that SEMA3E-plexin D1 and netrin-UNC5B might act at specific guide posts for developing vessels, thus suggesting that different repulsive signals might act at distinct vessel branching sites.

References

Adams RH, Eichmann A: Axon guidance molecules in vascular patterning, *Cold Spring Harb Perspect Biol* 2, 2010, a001875.

Basile JR, Barac A, Zhu T, et al.: Class IV semaphorins promote angiogenesis by stimulating rho-inititated pathways through plexin-B, *Cancer Res* 64:5212–5224, 2004.

Bates D, Taylor GI, Minichiello J, et al.: Neurovascular congruence results from a shared patterning mechanism that utilizes semaphorin3A and neuropilin-1, *Dev Biol* 255:77–98, 2003.

Carmeliet P, Tessier-Lavigne M: Common mechanisms of nerve and blood vessel wiring, *Nature* 436:193–200, 2005.

Conrotto P, Valdembri D, Corso S, et al.: SemaD4 induces angiogenesis through met recruitment by plexin B1, *Blood* 105:4321–4329, 2005.

Dickson BJ, Keleman K: Netrins, *Curr Biol* 12:R154–R155, 2002.

Kessler O, Shraga-Heled N, Lange T, et al.: Semaphorin-3F is an inhibitor of tumor angiogenesis, *Cancer Res* 64:1008–1015, 2004.

Kim J, Oh WJ, Gaiano N, et al.: Semaphorin 3E-plexin-D1 signaling regulates VEGF function in developmental angiogenesis via a feedback mechanism, *Genes Dev* 25:1399–1411, 2011.

Lu X, Le Noble F, Yuan L, et al.: The netrin receptor UNC5b mediates guidance events controlling morphogenesis of the vascular system, *Nature* 432:179–186, 2004.

Meadows SM, Ratliff LA, Singh MK, et al.: Resolution of defective dorsal aortae patterning in Sema-3E-deficient mice occurs via angiogenic remodelling, *Dev Dyn* 242:580–590, 2013.

Miao HQ, Soker S, Feiner L, et al.: Neuropilin-1 mediates collapsin-1/semaphorin III inhibition of endothelial cell notility. Functional competition of collapsin-1 and vascular endothelial growth factor-165, *J Cell Biol* 146:233–242, 1999.

Neufeld G, Kessler O: The semaphorins: versatile mediators of tumor progression and tumor angiogenesis, *Nat Rev Cancer* 8:632–645, 2008.

Serini G, Valdembri D, Zanivan S, et al.: Class 3 semaphorins control vascular morphogenesis by inhibiting integrin function, *Nature* 424:391–397, 2003.

Toyofuku T, Zhang A, Kumonogoh A, et al.: Dual roles of SEMA6D in cardiac morphogenesis through region-specific association of its receptor, Plexin-A1, with off track and vascular endothelial growth factor receptor-2, *Genes Dev* 18:435–447, 2004.

Toyofuku T, Yabuki M, Kamej J, et al.: Semaphorin-4A, an activator for T-cell-mediated immunity, suppresses angiogenesis via plexin-D1, *EMBO J* 26:1373–1384, 2007.

CHAPTER 9

Lumen formation and perfusion

Many organs, including blood vessels, are composed of epithelial tubes with an epithelial surface lining the lumen, which transport vital fluids (Lubraski and Krasnow, 2003). The lumen of a blood vessel is essential for providing blood to any given tissue. In the vascular system, lumen formation involves a complex molecular mechanism composed of endothelial cell repulsion on the cell-cell contacts within the endothelial cell cords, junctional rearrangement, and endothelial cell shape change (Iruela-Arispe and Davis, 2009). After the vascular lumen has been established, blood starts to flow through the newly formed vessel.

Early observations in the 18th century demonstrated that capillaries send out sprouts, which extend until they meet and anastomose with other sprouts or capillaries into which a lumen advances. Two different methods of lumen formation have been discussed: cord hollowing (Billroth, 1856) and cell hollowing (Sabin, 1920). In the former, a slit-like lumen formation takes place between the facing endothelial cells. In the latter, the ensuing lumen is the result of coalescence of intracytoplasmic vesicles. Clark and Clark also contributed other details to our knowledge of angiogenesis including the development of capillary lumina and fusion of sprouts (Clark and Clark, 1939).

During the final steps of capillary development, endothelial cell migration comes to a halt and endothelial cells form a lumen and re-establish functional adherens junctions (Dejana, 1996). Adherens junction formation was shown to be associated with the inhibition of endothelial cell migration in monolayers. This process was shown to be mediated by vascular endothelial cell cadherin (VE-cadherin) (Lampugnani et al., 1995; Dejana, 1996), a cell adhesion molecule belonging to the cadherin protein family, exclusively expressed in endothelial cells and facilitating their homotypic interaction (Cavallaro et al., 2006; Labelle et al., 2008). VE-cadherin is strictly required for the polarization of endothelial cells in vitro and in vivo, and VE-cadherin-based junctions are subjected to continuous reorganization, which renders them highly dynamic and sensitive to extracellular stimuli (Strilic et al., 2009; Lampugnani et al., 2010). Moreover, VE-cadherin transduces intracellular

New Insights on the Development of the Vascular System
https://doi.org/10.1016/B978-0-323-90599-2.00009-X

actomyosin contractility in endothelial cell-cell junctions in vivo (Lagendijk et al., 2017) and participates in a force-generating process during endothelial cell rearrangements (Paatero et al., 2018).

The lumen of the mouse dorsal aorta is formed by organized changes in endothelial cell shape (Strilic et al., 2009), while an alternative mode has been proposed in which intercellular and intracellular vacuoles form and fuse along connected endothelial cells (Kamei et al., 2006; Blum et al., 2008). Intracellular vacuoles coalesce to create a functional lumen within the sprouting intersegmental vessels in developing zebrafish embryos (Kamei et al., 2006). Intracellular vacuolization is a rapid method for creating endothelial cell luminal spaces (Bayless and Davis, 2002; Bayless et al., 2000; Davis and Camarillo, 1996; Koh et al., 2008). During this process, collagen (Davis and Senger, 2005) represents a pro-morphogenetic factor, whereas basement membrane proteins are inhibitory for vascular morphogenesis (Davis and Senger, 2005, 2008). In accordance with ultrastructural observations in the rat, mesentery lumen formation occurs relatively late in endothelial cords and first develop at the proximal parts of sprouts, whereas the tips of long and pointed sprouts of a considerable length do not contain detectable lumina (Rhodin and Fujita, 1989).

During lumen formation, two functionally different phenotypes of endothelial cells are recognized. The first phenotype, represented by endothelial cells in mature blood vessels, is characterized by an apicobasal polarity and junction-mediated contact inhibition. The second phenotype is found in activated tissues and is characterized by the loss of apicobasal polarity and adherens junctions, a spindle-shaped morphology, and the ability for guided migration.

Beta-1 integrin binds to the extracellular matrix and initiates the process of endothelial cell polarization. Adequate levels of β1 integrin are necessary for expression and localization of the intrinsic polarity protein in Par3 and other cell-cell interaction proteins, including VE-cadherin and platelet endothelial cell adhesion molecule (PECAM)-1 (Zovein et al., 2010; Malan et al., 2010). Endothelial cell polarization often starts with delivery via exocytosis of de-adhesive apical glycoproteins, which accumulate on the apical cell-cell contact (Strilic et al., 2010), including CD34 sialomucins such as CD34 and podocalyxin (PODXL), on the cell-cell contact (Martin-Belmonte et al., 2007; Ferrari et al., 2008). During delivery of apical glycoproteins, the adherens junctions translocate to lateral positions, and the new junctions have properties of adherens and tight junctions, as they contain both VE-cadherin (characteristic of adherens junctions) and zonula occludens-1 (ZO-1) protein (characteristic of tight junctions) (Bazzoni and Dejana, 2004).

Formation of apical cell surfaces and electrostatic repulsion of negatively charged apical glycoproteins are sufficient for the initial de-adhesion of adjacent endothelial cells and for slit formation but are not sufficient for the development of a patent vascular lumen. Therefore, vascular endothelial growth factor-A (VEGF-A) induces cell shape changes that further separate the apical cell surfaces from each other (Strilic et al., 2009). After lumen formation, the lumen diameter of vessel sprouting increases, and hemodynamic stimuli caused by shear stress significantly contribute to increased diameter. Vascular lumen expansion is force-dependent and involves F-actin cytoskeleton and/or blood flow. Stabilization of the newly formed vessel and maintenance of the existing vasculature are late events in the angiogenic process. Blood flow is critically important for determining the vessel fate (vessels with a high flow widen, whereas vessels with a low flow regress (Le Noble et al., 2004)) although recent advances in vascular biology have strongly argued for the autonomous fate control achieved by blood vessels (Jones, 2011). Chen et al. (2012) demonstrated that lymphatic endothelial and vessel phenotypes are negatively regulated by blood flow and that changes in hemodynamic forces can reprogram lymphatic vessels to blood vessels in postnatal life.

Genetic factors prevail during early embryogenesis, and, despite the lack of blood flow and mural cell coverage (Strilic et al., 2009; Wiegreffe et al., 2009), endothelial cell proliferation directly affects lumen diameter (Graupera et al., 2008; Stenman et al., 2008). A key extracellular matrix protein required for vascular lumen formation and remodeling is fibronectin (Bazigou et al., 2009; Stenzel et al., 2011), which is required for proper aortic lumen formation in the early mouse embryo and for retinal angiogenesis in the newborn mouse (Stenzel et al., 2011; Yang et al., 1993).

References

Bayless KJ, Davis GE: The Cdc42 and Rac1 GTPase are required for capillary lumen formation in three-dimensional extracellular matrices, *J Cell Sci* 115:123–1136, 2002.

Bayless KJ, Salazar R, Davis GE: RGD-dependent vacuolation and lumen formation observed during endothelial cell morphogenesis in three-dimensional fibrin matrices involves the alpha (v) beta (3) and alpha (5)-beta (1) integrins, *Am J Pathol* 156:1673–1683, 2000.

Bazigou E, Xie S, Chen C, et al.: Integrin-alpha9 is required for fibronectin matrix assembly during lymphatic valve morphogenesis, *Dev Cell* 17:175–186, 2009.

Bazzoni G, Dejana E: Endothelial cell-to-cell junctions: molecular organization and role in vascular homeostasis, *Physiol Rev* 84:869–901, 2004.

Billroth T: *Untersuchungen über die Entwicklung der Blutgefässe, nebst Beobachtungen aus der königlichen chirurigischen Universitäts-Klinik zu Berlin*, Berlin, 1856, Georg Reimer.

Blum Y, Belting HG, Ellestsdottir E, et al.: Complex cell rearrangements during intersegmental vessel sprouting and vessel fusion in zebrafish embryo, *Dev Biol* 316:312–322, 2008.

Cavallaro U, Liebner S, Dejana E: Endothelial cadherins and tumor angiogenesis, *Exp Cell Res* 312:659–667, 2006.

Chen CY, Bertozzi C, Zou Z, et al.: Blood flow reprograms lymphatic vessels to blood vessels, *J Clin Invest* 122:2006–2017, 2012.

Clark ER, Clark EL: Microscopic observations on the growth of blood capillaries in the living mammal, *Am J Anat* 64:251–301, 1939.

Davis GE, Camarillo CW: An alpha2 beta1 integrin-dependent pinocytotic mechanism involving intracellular vacuole formation and coalescence regulates capillary lumen and tube formation in three-dimensional collagen matrix, *Exp Cell Res* 224:39–51, 1996.

Davis GE, Senger DR: Endothelial extracellular matrix: biosynthesis, remodeling, and functions during vascular morphogenesis and neovessel stabilization, *Circ Res* 97:1093–1107, 2005.

Davis GE, Senger DR: Extracellular matrix mediates a molecular balance between vascular morphogenesis and regression, *Curr Opin Hematol* 15:197–203, 2008.

Dejana E: Endothelial adherens junctions: implications in the control of vascular permeability and angiogenesis, *J Clin Invest* 98:1949–1953, 1996.

Ferrari A, Veligodskiy A, Berge U, et al.: ROCK-mediated contractility, tight junctions and channels contribute to the conversion of a preapical patch into apical surface during isochoric lumen initiation, *J Cell Sci* 121:3649–3663, 2008.

Graupera M, Guillermet-Guibert J, Foukas LC, et al.: Angiogenesis selectively requires the pp110a isoform of PI3K to control endothelial cell migration, *Nature* 453:662–666, 2008.

Iruela-Arispe ML, Davis GE: Cellular and molecular mechanisms of vascular lumen formation, *Dev Cell* 16:222–231, 2009.

Jones EAV: The initiation of blood flow and flow induced events in early vascular development, *Semin Cell Dev Biol* 22:1028–1035, 2011.

Kamei M, Saunders WB, Bayless KJ, et al.: Endothelial tubes assemble from intracellular vacuoles in vivo, *J Cell Biol* 181:847–858, 2006.

Koh W, Stratman AN, Sacharidou A, et al.: In vitro three dimensional collagen matrix models of endothelial lumen formation during vasculogenesis and angiogenesis, *Methods Enzymol* 443:83–101, 2008.

Labelle M, Schnittler HJ, Aust DE, et al.: Vascular endothelial cadherin promotes breast cancer progression via transforming growth factor beta signaling, *Cancer Res* 68:1388–1397, 2008.

Lagendijk AK, Gomez GA, Backs S, et al.: Live imaging in junctional tension upon VE-cadherin in zebrafish, *Nat Commun* 8:1402, 2017.

Lampugnani MG, Corada M, Caveda L, et al.: The molecular organization of endothelial cell to cell junctions: differential association of plakoglobin, beta-catenin, and alpha-catenin with vascular endothelial cadherin (VE-cadherin), *J Cell Biol* 129:203–217, 1995.

Lampugnani MG, Orsenigo F, Rudini N, et al.: CCM1 regulates vascular-lumen organization by inducing endothelial polarity, *J Cell Sci* 123:1073–1080, 2010.

Le Noble F, Moyon D, Pardanaud L, et al.: Flow regulates arterial-venous differentiation in the chick embryo yolk sac, *Development* 131:361–375, 2004.

Lubraski B, Krasnow MA: Tube morphogenesis: making and shaping biological tubes, *Cell* 112:19–28, 2003.

Malan D, Wenzel D, Schmidt A, et al.: Endothelial beta 1 integrins regulate sprouting and network formation during vascular delvelopment, *Development* 137:993–1002, 2010.

Martin-Belmonte F, Gassama A, Datta A, et al.: PTEN-mediated apical segregation of phosphoinositides controls epithelial morphogenesis through Cdc42, *Cell* 128:383–397, 2007.

Paatero J, Santeur L, Lee M, et al.: Junction-based lamellipodia drive endothelial cell rearrangements in vivo via a VE-cadherin-F-actin based oscillatory cell-cell interactions, *Nat Commun* 9:3545, 2018.

Rhodin JAG, Fujita H: Capillary growth in the mesentery of normal young rats. Intravital video and electron microscope analyses, *J Submicrosc Cytol Pathol* 21:1–34, 1989.

Sabin F: Studies on the origin of blood-vessels and of red-corpuscles as seen in the living blastoderm of chicks during the second day of incubation, *Contrib Embryol* 9:213–262, 1920.

Stenman JM, Rajagopal J, Carroll TJ, et al.: Canonical Wnt signaling regulates organ-specific assembly and differentiation of CNS vasculature, *Science* 322:1247–1250, 2008.

Stenzel D, Lundkvist A, Sauvaget D, et al.: Integrin-dependent and -independent functions of astrocytic fibronectin in retinal angiogenesis, *Development* 138:4451–4463, 2011.

Strilic B, Kucera T, Eglinger J, et al.: The molecular basis of vascular lumen formation in the developing mouse aorta, *Dev Cell* 17:505–515, 2009.

Strilic B, Eglinger J, Krieg M, et al.: Electrostatic cell surface repulsion initiates lumen formation in developing blood vessels, *Curr Biol* 20:2003–2009, 2010.

Wiegreffe C, Christ B, Huang R, Scaal M: Remodeling of aortic smooth muscle during avian embryonic development, *Dev Dyn* 238:624–631, 2009.

Yang JT, Rayburn H, Hynes RO: Embryonic mesodermal defects in alpha 5 integrin-deficient mice, *Development* 119:1093–1105, 1993.

Zovein AC, Luque A, Turlo KA, et al.: Beta 1 integrin establishes endothelial cell polarity and arteriolar lumen formation via a Par-3 dependent mechanism, *Dev Cell* 18:39–51, 2010.

CHAPTER 10

Network formation

Following assembly of primitive vessels in the early embryo, remodeling transforms the plexus into an organized network of arteries, capillaries, and veins. Network formation represents a crucial step in the angiogenic process and provides the growing tissue with a newly constituted apparatus of immature and rudimentary vascular channels. The development of a complex, interconnected meshwork of crude capillary tubules is the structural substrate upon which acts the fine-tuning process of vascular remodeling. Network formation originates from coalescence of sprouts or by intussusceptive microvascular growth (IMG). Other mechanisms involving sprout fusion, tip cell filopodia, and macrophages, participate in vascular network formation. A sprouting vessel, by fusing with a target sprout, acquires a functional lumen and becomes stable (Iruela-Arispe and Davis, 2009). Tip cell filopodia interacts to initiate junction formation and reinforce sprout formation (Bentley et al., 2009). Macrophages may also facilitate sprout fusion by bridging sprouts to their potential target (Fantin et al., 2010).

The gross anatomy of the vascular system is characterized by highly reproducible branching patterns, with major and secondary branches forming at precisely designed sites and with organ-specific vascular architecture.

References

Bentley K, Mariggi G, Gerhardt H, et al.: Tipping the balance: robustness of tip selection, migration and fusion in angiogenesis, *PLoS Comput Biol* 5, 2009, e1000549.

Fantin A, Vieira JM, Gestri G, et al.: Tissue macrophages act as cellular chaperone sfor vascular anastomosis downstream ofVEGF-mediated endothelial tip cell induction, *Blood* 116:829–840, 2010.

Iruela-Arispe ML, Davis GE: Cellular and molecular mechanisms of vascular lumen formation, *Dev Cell* 16:222–231, 2009.

New Insights on the Development of the Vascular System
https://doi.org/10.1016/B978-0-323-90599-2.00010-6

CHAPTER 11
Remodeling

The initial endothelial plexus generated by vascular sprouting consists of a homogeneous web of endothelial cell tubes and sacs. This plexus is created in excess, and the final adjustment of vascular density involves regression of unnecessary vessels through a process of vascular remodeling and pruning, which creates a more differentiated vascular network (Risau, 1987).

Remodeling was documented by Clark and Clark in 1939, in a rabbit ear chamber model, in which anastomosis or fusion of vessel sprouts was demonstrated, combined with degeneration of vessels and thickening and reduction of other vessels with variation in flow and perfusion (Clark and Clark, 1939).

Remodeling involves the growth of new vessels and the regression of others as well as changes in the diameter of vessel lumens and in vascular wall thickening. Some vessels may fuse to form a larger one, such as the fusion of the paired dorsal aortae, or they may establish new connections like coronary vessels, which connect to the aorta. Examples of regression include the regression of capillaries in prechondrogenic regions to allow differentiation of the cartilage (Hallmann et al., 1987) and the regression of the hyaloid vasculature to allow development of a vitreous body in the eye (Latker and Kuwabara, 1981), angiogenesis in the corpus luteum of the cycling ovary (Augustin et al., 1995), and the retinal vasculature, undergoing dramatic vascular remodeling during the formation of the mature vasculature (Benjamin et al., 1998).

Blood flow is essential for remodeling of the primary vascular plexus into arteries and veins. In 1918, Chapman demonstrated by surgical removing the heart of chick embryo before the onset of circulation that the peripheral vasculature formed, but failed to remodel, without blood flow and pressure (Chapman, 1918).

Remodeling determines the formation of large and small vessels, establishment of directional flow, association with mural cells (pericytes and smooth muscle cells), and adjustment of vascular density to meet the nutritional requirements of the surrounding tissue. An interesting event that precedes vascular remodeling is the phenomenon of capillary retraction,

New Insights on the Development of the Vascular System
https://doi.org/10.1016/B978-0-323-90599-2.00011-8

described for the first time by Clark (1918) and Clark and Clark (1939). According to these authors, retraction of sprouts is observable during the rapid growth phase of capillaries and usually associates with elongation of other capillaries in the immediate vicinity. Retraction of capillary sprouts, which is not synonymous with regression of capillaries, could be defined as an event by which endothelial cells are withdrawn from one capillary sprout to become incorporated either into the mother vessel or into adjacent sprouts that are elongating. Remodeling is a complex phenomenon, which requires a vast array of molecular signaling. One of the earliest identified events involved in vascular remodeling was the interaction between the receptor tyrosine kinase Tie-2 (Schnürch and Risau, 1993; Sato et al., 1993) and its ligand, Angiopoietin (Ang) (Davis and Camarillo, 1996). Mice lacking Tie-2 or Ang die between embryonic days 9.5 and 12.5, and the embryos show a persistent capillary plexus, reminiscent of a defect in vascular remodeling and angiogenesis (Sato et al., 1995; Suri et al., 1996). Ang-2 is expressed before vessel remodeling and functions as a destabilizing signal, antagonizing Tie-2 function in vascular maintenance (Scharpfenecker et al., 2005). Researchers have constantly looked for identification of useful models for studying vascular remodeling. The developing vitreous vasculature can be viewed as a good in vivo model of vascular remodeling, maturation, and regression since it has a finite lifespan encompassing all of these processes. Indeed, formation and maturation of the vitreous vasculature involves extensive initial vasculogenesis followed by combinations of vasculogenesis and angiogenesis events, which are accompanied by pruning and remodeling of the maturing vitreous vessel network (Lang and Bishop, 1993; Yang and Cepko, 1996; Mitchell et al., 1998; Ash and Overbeek, 2000). Tubedown-1 (tbdn-1) expression in endothelial cells becomes downregulated during the formation of capillary-like structures in vitro and is regulated in vivo in a manner that suggests a functional role in dampening blood vessel development. Paradis et al. (2002) showed that tbdn-1 was expressed highly in the vitreous vascular network during the pruning and remodeling phases of this transient structure. Overall, these data indicate that the developing vasculature responds dynamically to the growing needs of the embryo by remodeling vessels as required.

It has been recognized that endothelial cells use other cells as migration tracks during capillary network remodeling. In addition to sprouting, Nehls et al. (1998) indeed defined a new type of endothelial cell migration, namely, guided migration of endothelial cells along performed capillary-like structures. By fluorescent labeling of living endothelial cells, these authors observed that endothelial cells in immature microvascular networks migrated along preformed capillary-like structures. By guided migration, endothelial cells increase on preexisting capillary-like structures, which consequently lead to thickening and to increased capillary diameters. Endothelial cell-derived fibroblast growth factor-2 (FGF-2) is involved in the regulation of guided migration.

Vascular remodeling is a key promoter of intratumoral perfusion and oxygenation, preventing tumor invasiveness and metastasis (Van Beijnum et al., 2015).

References

Ash JD, Overbeek PA: Lens-specific VEGF-A expression induces angioblast migration and proliferation and stimulates angiogenic remodeling, *Dev Biol* 223:383–398, 2000.

Augustin HG, Braun K, Telemenakis I, et al.: Ovarian angiogenesis. Phenotypic characterization of endothelial cells in a physiological model of blood vessel growth anf regression, *Am J Pathol* 147:339–351, 1995.

Benjamin LE, Hemo I, Keshet E: A plasticity window for blood vessel remodeling is defined by pericyte coverage of the preformed endothelial network and is regulated by PDGF-B and VEGF, *Development* 125:1591–1598, 1998.

Chapman WB: The effect of the heart-beat upon the development of the vascular system in the chick, *Am J Anat* 23:175–203, 1918.

Clark ER: Studies on the growth of blood vessels in the tail of the frog larva by observation and experiment on the living animal, *Am J Anat* 23:37–88, 1918.

Clark ER, Clark EL: Microscopic observations on the growth of blood capillaries in the living mammal, *Am J Anat* 64:251–301, 1939.

Davis GE, Camarillo CW: An alpha2 beta1 integrin-dependent pinocytotic mechanism involving intracellular vacuole formation and coalescence regulates capillary lumen and tube formation in three-dimensional collagen matrix, *Exp Cell Res* 224:39–51, 1996.

Hallmann R, Feinberg RN, Latker CH, et al.: Regression of blood vessels precedes cartilage differentiation during chick limb development, *Differentiation* 34:98–105, 1987.

Lang RA, Bishop JM: Macrophages are required for cell death and tissue remodeling in the developing mouse eye, *Cell* 74:453–462, 1993.

Latker CH, Kuwabara T: Regression of tunica vasculosa lentis in the postnatal rat, *Invest Ophthalmol Vis Sci* 21:689–699, 1981.

Mitchell CA, Risau W, Drexler HC: Regression of vessels in the tunica vasculosa lentis is initiated by coordinated endothelial apoptosis: a role for vascular endothelial growth factor as a survival factor for endothelium, *Dev Dyn* 213:322–333, 1998.

Nehls V, Hermann R, Huhnken M: Guided migration as a novel mechanism of capillary network remodeling is regulated by basic fibroblast growth factor, *Histochem Cell Biol* 109:309–319, 1998.

Paradis H, Liu CY, Saika S, et al.: Tubedown-1 in remodeling of the developing vitreal vasculature in vivo and regulation of capillary outgrowth in vitro, *Dev Biol* 249:140–155, 2002.

Risau W: Mechanisms of angiogenesis, *Nature* 386:671–674, 1987.

Sato TN, Qin Y, Kozak CA, et al.: Tie-1 and Tie-2 define another class of putative receptor tyrosine kinase genes expressed in early embryonic vascular system, *Proc Natl Acad Sci U S A* 90:9355–9358, 1993.

Sato TN, Tozawa Y, Deutsch U, et al.: Distinct roles of the receptor tyrosine kinase Tie-1 and Tie-2 in blood vessel formation, *Nature* 376:70–74, 1995.

Scharpfenecker M, Fiedler U, Reiss Y, et al.: The Tie-2 ligand angiopoietin-2 destabilizes quiescent endothelium through an inetranal autocrine loop mechanism, *J Cell Sci* 118:771–780, 2005.

Schnürch H, Risau W: Expression of tie-2, a member of a novel family of receptor tyrosine kinases, in the endothelial cell lineage, *Development* 119:957–968, 1993.

Suri C, Jones PF, Ptan S, et al.: Requisite role of angiopoietin-1, a ligand for the TIE2 receptor, during embryonic angiogenesis, *Cell* 87:1171–1180, 1996.

Van Beijnum J, Nowak-Sliwinska P, Huijbers EJM, et al.: The great escape: the hallmarks of resistance to antiangiogenic therapy, *Pharmacol Rev* 67:441–461, 2015.

Yang K, Cepko CL: Flk-1, a receptor for vascular endothelial growth factor (VEGF), is expressed by retinal progenitor cells, *J Neurosci* 16:6089–6099, 1996.

CHAPTER 12
Pruning

Pruning is a mechanism that allows the vascular system to adapt to the changing hemodynamic and metabolic influences and to create a more efficient angioarchitecture, thus decreasing vascular complexity and increasing vascular diameter (Udan et al., 2013). In response to blood flow, endothelial cells migrate from low-flow/no-flow segments toward high-flow segments, leading to regression of low-flow segments and stabilization of high-flow segments (Fig. 1) (Franco et al., 2015).

Pruning was first described in the embryonic retina and involves the removal of excess of endothelial cells, which form redundant channels (Ashton, 1966). Simulations suggest how flow might affect vessel pruning, which, due to low wall shear stress, is highly sensitive to the pressure drop across the vascular network. In fact, the degree of pruning increases as the pressure drop increases (Skalak and Price, 1996). Pruning and remodeling of the vascular network may be stimulated by tissue-derived signaling molecules and blood flow conditions (e.g., wall shear stress and pressure). Although the morphological hallmarks of intussusceptive microvascular growth (IMG) and intussusceptive arborization (IAR), which are crucial mechanisms for generating and expanding a capillary network, were frequently observed in the capillaries and terminal microvessels with diameters $<25\,\mu m$, the formation of transluminal pillars and folds in the arteries and veins with diameters of up to $110\,\mu m$ was a surprise. Such pillars were closely associated with bifurcations and were shown to be involved in remodeling and pruning of larger microvessels (Djonov et al., 2002). Hence, this process was termed intussusceptive branching remodeling (IBR).

An interesting model of vascular pruning is provided by modification of pulmonary vascularization under certain conditions (Howell et al., 2003). Chronic hypoxia caused by migration of native sea-level dwellers to higher altitudes or chronic lung disease leads to the development of increased pulmonary vascular resistance and pulmonary hypertension (Meyrick and Reid, 1978; Hopkins and McLoughlin, 2002). Structural changes that are believed to underlie the increased vascular resistance can be broadly classified into two processes: first, remodeling of the walls of the pulmonary

New Insights on the Development of the Vascular System
https://doi.org/10.1016/B978-0-323-90599-2.00012-X

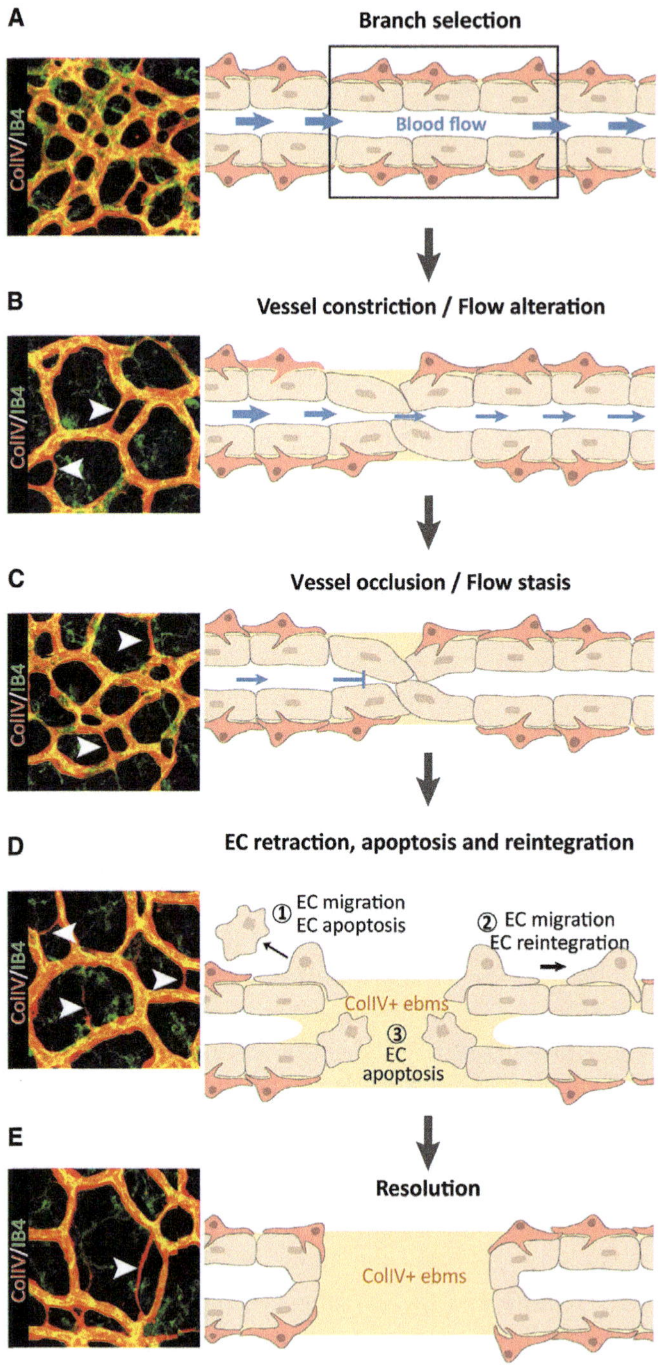

Fig. 1 See the legend in facing page

resistance vessels and second pruning in the total number of blood vessels in the lung (Meyrick and Reid, 1978; Rabinovitch et al., 1979). The structural changes include muscularization of nonmuscular arterioles, increased medial thickness of muscular arterioles, adventitial hypertrophy, and deposition of additional matrix components, including collagen and elastin, in the vascular walls (Rabinovitch et al., 1979; Stenmark and Mecham, 1997). The second major structural alteration caused by chronic hypoxia is loss of small blood vessels, which is said to increase vascular resistance by reducing the extent of parallel vascular pathways (Hislop and Reid, 1976, 1977; Rabinovitch et al., 1979).

In another vascular system, the retina, oxygen supply is regarded as a key factor in blood vessel pruning. Earlier studies have highlighted the key role of oxygen and vascular endothelial growth factor (VEGF) in the formation and remodeling of the retina vasculature (Stone and Maslim, 1997; Dorrell and Friedlander, 2006). In addition to preventing new vessel growth, hyperoxia suppresses VEGF production and leads to obliteration by apoptosis of the already formed vessels (Stone et al., 1995). This suggests that vascular pruning accompanying natural remodeling is caused by hyperoxia upon the onset of flow through the newly formed vascular system. Exposure to hyperoxia leads to excessive regression of the capillaries, while the arteries become refractory to this insult (Benjamin et al., 1998). Excessive vascular pruning is the initiating event in the pathogenesis of retinopathy of prematurity (ROP), a blindness-causing disease induced in premature infants placed in oxygen chambers, and VEGF protects retinal vessels from hyperoxia-induced obliteration in an experimental model of ROP (Alon et al., 1995). Activated leukocytes extravasate in the retina and are involved in mediation of endothelial cell apoptosis (Ishida et al., 2003).

Fig. 1 Vessel pruning and regression processes. (A–D) Following the primarily blood flow-driven selection of a branch for regression (A), the vessel constricts (B) until it occludes (C) and the blood flow ceases (D). Endothelial cells (ECs) in regressing vessel segments may retract and undergo apoptosis (1), or they may migrate away to reintegrate elsewhere (2), leaving behind collagen IV (collV) + empty basement membrane sleeves (ebms). Retracting endothelial cells might disintegrate from the vascular network and undergo endothelial cell apoptosis due to detachment from the basement membrane (3). (E) In the final resolution step, regression of the selected branch is completed, leaving behind a remodeled vascular network. (Reproduced from Korn C, Augustin HG: Mechanisms of vessel pruning and regression, *Dev Cell* 34:5–17, 2015.)

References

Alon T, Hemo I, Itin A, et al.: Vascular endothelial growth factor acts as a survival factor for newly formed retinal vessels and has implications for retinopathy of prematurity, *Nat Med* 1:1024–1208, 1995.

Ashton N: Oxygen and the growth and development of retinal vessels, *Am J Ophthalmol* 62:412–435, 1966.

Benjamin LE, Hemo I, Keshet E: A plasticity window for blood vessel remodeling is defined by pericyte coverage of the preformed endothelial network and is regulated by PDGF-B and VEGF, *Development* 125:1591–1598, 1998.

Djonov VG, Kurz H, Burri PH: Optimality in the developing vascular system: bifurcation remodeling by means of intussusception as an efficient adaptation mechanism, *Dev Dyn* 224:391–402, 2002.

Dorrell MI, Friedlander M: Mechanisms of endothelial cell guidance and vascular patterning in the developing mouse retina, *Prog Retin Eye Res* 25:277–295, 2006.

Franco CA, Jones ML, Bernabeu MO, et al.: Dynamic endothelial cell rearrangements drive developmental vessel regression, *PLoS Biol* 13, 2015, e1002125.

Hislop A, Reid L: New findings in pulmonary arteries of rats with hypoxia-induced pulmonary hypertension, *Br J Exp Pathol* 57:542–554, 1976.

Hislop A, Reid L: Changes in the pulmonary arteries of the rat during recovery from hypoxia-induced pulmonary hypertension, *Br J Exp Pathol* 58:653–662, 1977.

Hopkins N, McLoughlin P: The structural basis of pulmonary hypertension in chronic lung disease: remodeling, rarefaction or angiogenesis, *J Anat* 201:335–348, 2002.

Howell K, Preston RJ, McLoughlin P: Chronic hypoxia causes angiogenesis in addition to remodeling in the adult pulmonary circulation, *J Physiol* 547:133–145, 2003.

Ishida S, Yamashiro K, Usui T, et al.: Leukocytes mediate retinal vascular remodeling during development and vaso-obliteration in disease, *Nat Med* 9:781–788, 2003.

Meyrick B, Reid L: The effect of continued hypoxia on rat pulmonary arterial circulation. An ultrastructural study, *Lab Investig* 38:188–200, 1978.

Rabinovitch M, Gamble W, Nadas AS, et al.: Rat pulmonary circulation after chronic hypoxia: hemodynamic and structural features, *Am J Phys* 236:H818–H827, 1979.

Skalak TC, Price RJ: The role of mechanical stresses in microvascular remodeling, *Microcirculation* 3:143–165, 1996.

Stenmark KR, Mecham RP: Cellular and molecular mechanisms of pulmonary vascular remodeling, *Annu Rev Physiol* 59:89–144, 1997.

Stone J, Maslim J: Mechanisms of retinal angiogenesis, *Prog Retin Eye Res* 16:157–181, 1997.

Stone J, Itin A, Alon T, et al.: Development of retinal vasculature is mediated by hypoxia-induced vascular endothelial growth factor (VEGF) expression by neuroglia, *J Neurosci* 15:4783–4787, 1995.

Udan RS, Vadakkan TJ, Dickinson ME: Dynamic responses of endothelial cells to changes in blood flow during vascular remodelling of the mouse yolk sac, *Development* 40:4041–4050, 2013.

CHAPTER 13

Maturation and stabilization and pericyte recruitment

Several cellular and noncellular components in the blood vessel, including endothelial cells, pericytes, smooth muscle cells, fibroblasts, glial cells, inflammatory cells, and the extracellular matrix, coordinately regulate the maintenance of vessel integrity at varying degrees in different vascular beds (Xu and Cleaver, 2011; Patel-Hett and D'Amore, 2012; Stratman and Davis, 2012).

Pericytes are recruited by differentiation from surrounding mesenchymal precursors or by migration from the mural wall of the adjacent vessel (Gerhardt and Betsholtz, 2003). In this manner, pericytes suppress endothelial growth (Orlidge and D'Amore, 1987) and migration (Sato and Rifkin, 1989). There is a striking coincidence between pericyte investment and microvessel stabilization (Von Tell et al., 2006; Bergers and Song, 2005), and pericyte investment has also been directly implicated in conferring capillary resistance to regression in vivo (Benjamin et al., 1998). Clinical evidence for a stabilizing function of pericytes was provided by the finding that the development of microaneurysms of retinal capillaries, as a symptom of diabetic retinopathy, was correlated with an initial loss of intramural pericytes (Kuwabara and Cogan, 1963).

Stratman et al. (2009) demonstrated that endothelial cell-pericyte interactions regulate increased expression of basement membrane protein genes and proteins, such as fibronectin and laminin, as well as integrins that recognize the remodeled matrices to control this process and these changes occur specifically in endothelial cell-pericyte cocultures and not in endothelial cell-only cultures. Pericytes can invade tissues in the absence of endothelial cells and can form tubes, thus enabling the subsequent penetration of endothelial cells (Ozerdem and Stallcup, 2003). Rajantie et al. (2004) showed that bone marrow-derived hematopoietic progenitor cells CD11b-positive and CD45-positive expressing the pericyte marker nerve/glial antigen-2 (NG2) were located in close proximity to blood vessels in a subcutaneous B16-F1 melanoma model. Bone marrow-derived platelet derived growth factor

New Insights on the Development of the Vascular System
https://doi.org/10.1016/B978-0-323-90599-2.00013-1

receptor beta (PDGFR-β)-positive/Sca-1-positive progenitor pericytes have been demonstrated in a mouse model of pancreatic islet tumorigenesis, which were able to differentiate into mature pericytes expressing the markers NG2 and alpha smooth actin (Song et al., 2005). Virgintino et al. (2007) demonstrated in the human fetal telencephalon that growing microvessels are formed by a pericyte-driven angiogenic process in which endothelial cells are preceded and guided by migrating pericytes. Endothelial cells direct the differentiation of mural cell precursors from the adjacent tissue by the secretion of soluble factors (Beck Jr and D'Amore, 1997; Nehls et al., 1992). A different view is that mural cells become associated with endothelial cells by migrating along newly developed vascular sprouts (Nicosia and Villaschi, 1995).

A role for mural cells in maintaining vascular integrity was suggested by a number of gene knockout studies, including disruption of the genes encoding Tie-1 and Tie-2 (Dumont et al., 1994; Puri et al., 1995), Angiopoietin-1 (Ang-1) (Suri et al., 1996), tissue factor (TF) system (Carmeliet et al., 1996), and PDGF-B/PDGFR-β[a] (Leveen et al., 1994; Lindahl et al., 1997). Mice deficient in these genes show a hemorrhaging phenotype often associated with a reduced number of alpha-smooth muscle actin (α-SMA)-positive perivascular cells and microaneurysm formation. Ang-1- and TF-deficient mice also show defects in the remodeling of the homogeneous yolk sac vascular plexus.

Endothelial cells secrete PDGF-B and pericytes express PDGFR-β (Betsholtz, 2004), suggesting a paracrine mode of interaction between these two cell types (Fig. 1). PDGF-B promotes pericyte precursor cell proliferation and migration and mice deficient in PDGF-B or PDGFR-β die during embryonic development with widespread microvascular defects, consisting of vessel dilatation and microaneurysms, altered organization of endothelial cell junctions, increased vesicular transport, increased leakage of plasma and blood cells, and drastic reduction of pericyte association with vessels (Betsholtz, 2004). Endothelial cells of the sprouting capillaries in PDGF-B-deficient mice were unable to attract PDBFR-β-positive progenitors of pericytes, and failure to recruit pericytes during development leads to vascular instability and regression (Benjamin et al., 1998; Leveen et al., 1994; Lindblom et al., 2003). Under conditions of

[a] The PDGF family of growth factors is encoded by four genes that give rise to four homdimeric proteins, namely, PDGF-AA, -BB, -CC, and -DD, and one heterodimeric form, PDGF-AB. The two PDGF receptors are designated α and β receptors. The five PDGF ligands display distinct receptor-binding specificities. PDGF-BB induces the formation of all three possible receptor dimers (α/α, α/β, and β/β), whereas PDGF-AA and -DD signal only through α/α and β/β receptors, respectively. Both PDGF-AB and -CC can induce receptor heterodimers and, in addition, signal through PDGF-α receptor homodimers.

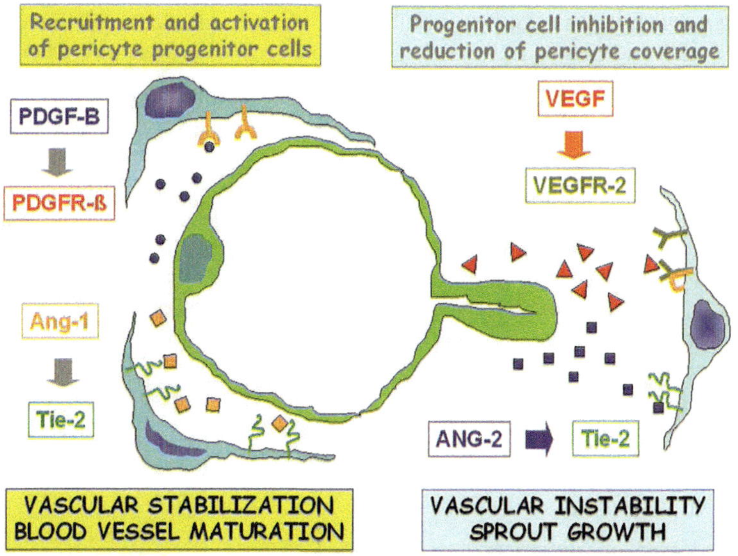

Fig. 1 Schematic of the paracrine interactions occurring between pericyte precursor cells and endothelial cells in PDGF-mediated angiogenesis. Endothelial cells secrete PDGF-B, which causes pericyte precursor cell proliferation and migration through activation of PDGFR-β receptors. Pericytes surround and cover early endothelial tubes. In contrast, endothelial cells in vascular sprouts release VEGF, which in turn mediates suppression of PDGFR-β signaling through the induction of VEGFR-2/PDGFR-β complexes. This pathway abrogates pericyte coverage of endothelial sprouts, leading to vascular instability and regression. (Reproduced from Ribatti D, Nico B, Crivellato E: The role of pericytes in angiogenesis, *Int J Dev Biol* 55:261–268, 2011.)

PDGF-mediated angiogenesis, VEGF abolished pericyte coverage of vascular sprouts, leading to vessel destabilization, and VEGF-mediated activation ofVEGF receptor-2 (VEGFR-2) suppressed PDGFR-β signaling, through the induction of a VEGFR-2/PDGFR-β complex (Fig. 2) (Greenberg et al., 2008).

Transforming growth factor-beta 1 (TGF-β1) appears to be instrumental in the de novo induction of pericytes by regulating differentiation of pericyte progenitors. When mesenchymal cells are cocultured with endothelial cells or treated with TGF-β1, they express smooth muscle cell markers, indicating differentiation of precursor cells into pericytes or smooth muscle cells (Darland and D'Amore, 2001). Mice deficient in endoglin, a TGF-β1 coreceptor, display reduced association with smooth muscle cells and pericytes (Li et al., 1999). TGF-β1 inhibits endothelial cell proliferation and migration, and mice deficient in TGF-β1 signaling components show dilated and irregularly shaped microvessels (Lebrin et al., 2005).

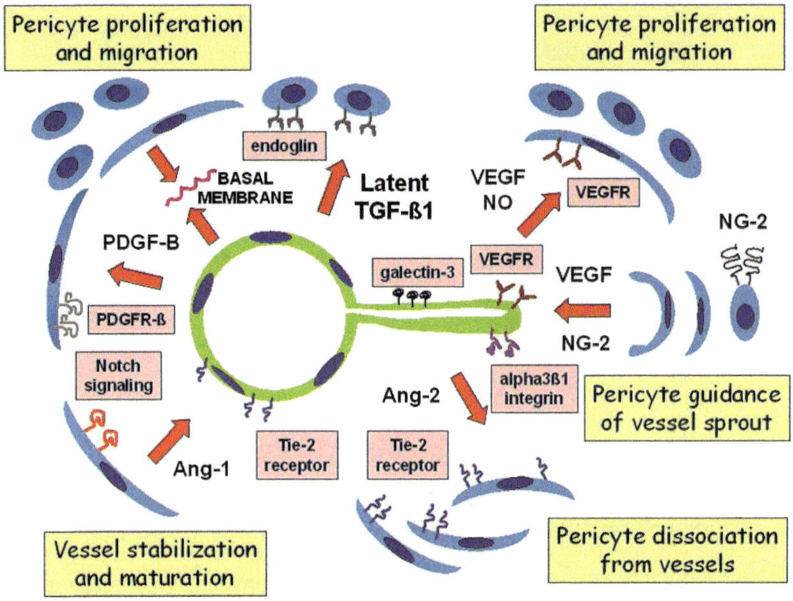

Fig. 2 Signaling pathways operating in endothelial cell/pericyte paracrine crosstalk. Pericytes are involved in endothelial cell stimulation and guidance as well as in endothelial stabilization and maturation. Vessel sprouts (right) cause destabilization of pericyte investment through Ang-2/Tie-2 signaling. Pericytes provide guidance for endothelial movement and tube formation through secretion of VEGF and soluble NG2. Spreading endothelial cells, in turn, stimulate pericyte precursor cell proliferation and migration by releasing VEGF and NO. Vessel stabilization (left) occurs by pericyte investment and close interaction with endothelial cells. Mature endothelial cells secrete PDGF-B, which promotes proliferation and migration of pericyte precursor cells through activation of PDGFR-β receptors expressed on the surface of pericyte progenitors. This mechanism leads to pericyte coverage of early endothelial tubes. Vessel maturation further develops through Ang-1- and Notch-mediated signaling. Pericyte stabilize and reinforce the endothelial tube, contributing to secretion of the basement membrane. (Reproduced from Ribatti D, Nico B, Crivellato E: The role of pericytes in angiogenesis, *Int J Dev Biol* 55:261–268, 2011.)

VEGF directly induces proliferation and migration of pericytes under hypoxic conditions (Yamagishi et al., 1999) and also indirectly stimulates pericyte recruitment via endothelial cell production of nitric oxide (NO). In turn, NO promotes mural precursor cell migration in vitro and pericyte recruitment to tumor vessels in vivo (Kashiwagi et al., 2005). Treatment with VEGF inhibitors causes pericytes to become closely associated with the surviving tumor vessels in Lewis lung carcinomas, RIP-Tag2 tumors (Inai et al., 2004), and other tumor models (Tong et al., 2004; Willet et al., 2004). Darland et al. (2003) demonstrated that pericytes cocultured with endothelial cells produce VEGF that may act in a juxtacrine/paracrine manner as a survival and/or a stabilizing factor for endothelial cells. Moreover, they observed VEGF gene expression in the developing retinal vasculature in pericytes contacting newly formed microvessels.

Ang-1 is produced by pericytes and smooth muscle cells, activates endothelial Tie-2, maximizes interactions between endothelial cells and pericytes, and is expressed behind the leading edge of angiogenic vessels (Sundberg et al., 2002). Mice deficient in either Ang-1 or Tie-2 die during embryonic development with vascular defects similar to those observed for PDGF-B-deficient mice (Jones et al., 2001). Ultrastructural analysis suggests that Tie-2 knockout blood vessels lack mural cells (Patan, 1998). In PDGF-B-deficient mice, recombinant Ang-1 restored the vascular structure and permeability in the growing retinal vasculature (Uemura et al., 2002), and Ang-1 also counteracted VEGF-induced endothelial leakiness (Thurston et al., 1999). Knockout embryos lacking Ang-1 display failure of endothelial cell adherence and interaction with perivascular cells and the extracellular matrix (Davis and Yancopoulos, 1999). In the absence of mural cells, recombinant Ang-1 restored a hierarchical order of the larger vessels and rescued edema and hemorrhage in the growing retinal vasculature of mouse neonates (Uemura et al., 2002).

The expression of both Ang-2 and Tie-2 in pericytes has been reported (Wakui et al., 2006; Cai et al., 2008). VEGF increases production of Ang-2 and overexpression of Ang-2, which binds to Tie-2 in competition with Ang-1 by endothelial cells, and results in dissociation of pericytes from vessels (Zhang et al., 2003), reduction of pericyte coverage, and destabilization of vessels within the tumor even in the presence of VEGF stimulation (Cao et al., 2007). Moreover, transgenic mice overexpressing Ang-2 in the retina develop dense vascular networks with reduced pericyte coverage (Feng et al., 2007). Immature pericytes express the NG2 proteoglycan during early stages of angiogenesis, and soluble NG2 promotes endothelial cell motility and angiogenesis forming a complex with galectin-3 and $\alpha3\beta1$ integrin on the cell surface (Fukushi et al., 2004). Both blocking by antibodies and knocking out of the gene encoding NG2 abrogated vascular growth (Ozerdem and Stallcup, 2004). Virgintino et al. (2007) showed that microvessels of the fetal human telencephalon are characterized by a continuous layer of NG2-positive pericytes, which tightly invest endothelial cells in the earliest stages of vessel growth.

The Notch-3 receptor is highly expressed in pericytes, and disruption of Notch-3 signaling in Notch-3$^{-/-}$ mutant mice results in enlarged vessels due to the lack of pericytes (Wang et al., 2007). Patients suffering from the

CADASIL (cerebral autosomal dominant arteriopathy with subcortical infarcts and leukoencephalopathy) syndrome, pathology associated with mutations of Notch-3, present vessels lacking pericytes (Louvi et al., 2006). In the dental pulp, perivascular cells mainly express Notch-3 (Lovschall et al., 2007), and, in the retina, pericytes also express Notch-3 (Claxton and Fruttiger, 2004). Liu et al. (2009) showed that knock down by small interfering RNA revealed that Notch-3 signaling is necessary for endothelial-dependent mural cell differentiation and that Notch-3 contributes to the proangiogenic capability of mural cells cocultured with endothelial cells.

Ephrin-B (Eph-B) is a critical mediator of postnatal pericyte-to-endothelial cell assembly into vascular tubes. Furthermore, inhibition of Eph-B activity prevents assembly of pericytes and endothelial cells (Salvucci et al., 2009).

During the initial phase of angiogenesis, activated pericytes in parent vessels bulge, shorten their processes and increase their volume, an intense cell proliferation occurs, pericytes project into the perivascular spaces, and their basement membrane is disrupted and fragmented and appear detached from the vessel wall (Diaz-Flores et al., 1992). Although initially endothelial cell sprouts may form without pericyte involvement, pericytes are among the first cells to invade the newly vascularized tissues and locate at the growing front of the endothelial sprouts by determining the location of sprout formation and by guiding the newly formed vessels (Nehls et al., 1992). Individual pericytes can be found at the tips of angiogenic sprouts in the corpus luteum, where pericytes are the first vascular cells to invade the granulose fold of the ruptured follicle, and in tumors (Amselgruber et al., 1999; Gerhardt and Betsholtz, 2003; Morikawa et al., 2002).

In the ocular model, changes in the state of ocular pericytes are associated with vasculopathy in eye diseases in humans. Pericyte death, thickening of the basement membrane, and changes in extracellular matrix expression pattern in proliferative diabetic retinopathy have been described (Yonoff, 1966; Bloodworth Jr and Epstein, 1967; Cogan and Kuwalara, 1967; Addison et al., 1970; Ashton, 1974; Speiser et al., 1968; Podestà et al., 2000). By studying postnatal remodeling of the retina vasculature, it has been observed that pericyte recruitment represented a critical step in vascular maturation (Benjamin et al., 1998). It is preceded by outmigration of cells positive for α-SMA from the arterioles, and coverage of primary and smaller branches lagged many days behind formation of the endothelial plexus. The transient existence of a pericyte-free endothelial plexus coincided temporally and spatially with the process of hyperoxia-induced vascular pruning, which is a mechanism for fine-tuning of vascular density according to available oxygen (Benjamin et al., 1998).

References

Addison DJ, Garner A, Ashton N: Degeneration of intramural pericytes un diabetic retinopathy, *Br Med J* 5691:264–266, 1970.

Amselgruber WM, Schafer M, Sinowatz F: Angiogenesis in the bovine corpus luteum: an immunohistochemical and ultrastructural study, *Anat Histol Embryol* 28:157–166, 1999.

Ashton N: Vascular basement membrane changes in diabetic retinopathy. Montgomery lecture, 1973, *Br J Ophthalmol* 58:344–366, 1974.

Beck L Jr, D'Amore PA: Vascular development: cellular and molecular regulation, *FASEB J* 11:365–373, 1997.

Benjamin LE, Hemo I, Keshet E: A plasticity window for blood vessel remodeling is defined by pericyte coverage of the preformed endothelial network and is regulated by PDGF-B and VEGF, *Development* 125:1591–1598, 1998.

Bergers G, Song S: The role of pericytes in blood-vessel formation and maintenance, *Neuro-Oncology* 7:452–464, 2005.

Betsholtz C: Insight into the physiological functions of PDGF through genetic studies in mice, *Cytokine Growth Factor Rev* 15:215–228, 2004.

Bloodworth JM Jr, Epstein M: Diabetic amyotrophy. Light and electron microscopy investigation, *Diabetes* 16:181–190, 1967.

Cai J, Kehoe O, Smith GM, et al.: The angiopoietin/Tie-2 system regulates pericyte survival and recruitment in diabetic retinopathy, *Invest Ophthalmol Vis Sci* 49:2163–2171, 2008.

Cao Y, Sonveaux P, Liu S, et al.: Systemic overexpression of angiopoietin-2 promotes tumor microvessel regression and inhibits angiogenesis and tumor growth, *Cancer Res* 67:3835–3844, 2007.

Carmeliet P, Mackman N, Moons L, et al.: Role of tissue factor in embryonic blood vessel development, *Nature* 383:73–75, 1996.

Claxton S, Fruttiger M: Periodic Delta-like 4 expression in developing retinal arteries, *Gene Expr Patterns* 5:123–127, 2004.

Cogan DG, Kuwalara T: The mural cell in perspective, *Arch Ophthalmol* 78:133–139, 1967.

Darland DC, D'Amore PA: Cell-cell interactions in vascular development, *Curr Top Dev Biol* 52:107–149, 2001.

Darland DC, Massingham LJ, Smith SR, et al.: Pericyte production of cell-associated VEGF is differentiation-dependent and is associated with endothelial survival, *Dev Biol* 264:275–288, 2003.

Davis S, Yancopoulos GD: The angiopoietins: yin and Yang in angiogenesis, *Curr Top Microbiol Immunol* 237:173–185, 1999.

Diaz-Flores L, Gutierrez R, Varela H: Behavior and lineage of postcapillary venule pericytes during postnatal angiogenesis, *J Morphol* 213:33–45, 1992.

Dumont DJ, Gradwohl G, Fong GH, et al.: Dominant-negative and targeted null mutations in the endothelial receptor tyrosine kinase, tek, reveal a critical role in vasculogenesis of the embryo, *Genes Dev* 8:1897–1909, 1994.

Feng Y, Vom Hagen F, Pfister F, et al.: Impaired pericyte recruitment and abnormal retinal angiogenesis as a result of angiopoietin-2 overexpression, *Thromb Haemost* 97:99–108, 2007.

Fukushi J, Makangiansar IT, Stallcup WB: NG2 proteoglycan promotes endothelial cell motility and angiogenesis via engagement of galectin-3 and alpha3 beta1 integrin, *Mol Biol Cell* 15:3580–3590, 2004.

Gerhardt H, Betsholtz C: Endothelial-pericyte interactions in angiogenesis, *Cell Tissue Res* 314:15–23, 2003.

Greenberg JI, Shields DJ, Barillas SG, et al.: A role for VEGF as a negative regulator of pericyte function and vessel maturation, *Nature* 456:809–813, 2008.

Inai T, Mancuso M, Hashizume H, et al.: Inhibition of vascular endothelial growth factor (VEGF) signaling in cancer causes loss of endothelial fenestrations, regression of tumor vessels, and appearance of basement membrane ghosts, *Am J Pathol* 165:35–52, 2004.

Jones N, Iliin K, Dumont DJ, et al.: Tie receptors: new modulators of angiogenic and lymph-angiogenic responses, *Nat Rev Mol Cell Biol* 2:257–267, 2001.

Kashiwagi S, Izumi Y, Gohongi T, et al.: NO mediates mural cell recruitment and vessel morphogenesis in murine melanomas and tissue-engineered blood vessels, *J Clin Invest* 115:181618–181627, 2005.

Kuwabara T, Cogan DG: Retinal vascular patterns. VI. Mural cells of the retinal capillaries, *Arch Ophthalmol* 69:492–502, 1963.

Lebrin F, Deckers M, Bertolino P, et al.: TGF-beta receptor function in the endothelium, *Cardiovasc Res* 65:599–608, 2005.

Leveen P, Pekny M, Gebre Medhin S, et al.: Mice deficient for PDGF B show renal, cardiovascular, and hematological abnormalities, *Genes Dev* 8:1875–1887, 1994.

Li DY, Sorensen LK, Brooke BS, et al.: Defective angiogenesis in mice lacking endoglin, *Science* 284:1534–1537, 1999.

Lindahl P, Johansson BR, Levéen P, et al.: Pericyte loss and microaneurysm formation in PDGF-B-deficient mice, *Science* 277:242–245, 1997.

Lindblom P, Gerhardt H, Liebner S, et al.: Endothelial PDGF-B retention is required for proper investment of pericyte in the microvessel wall, *Genes Dev* 17:1835–1840, 2003.

Liu H, Kennards S, Lilly B: NOTCH3 expression is induced in mural cells through an autoregulatory loop that requires endothelial-expressed JAGGED1, *Circ Res* 104:466–475, 2009.

Louvi A, Arboleda-Velasquez JF, Artavans-Tsakonas S: CADASIL: a critical look at a notch disease, *Dev Neurosci* 28:5–12, 2006.

Lovschall H, Mitsiadist A, Poulsen K, et al.: Coexpression of Notch3 and Rgs5 in the pericyte-vascular smooth muscle cell axis in response to pulp injury, *Int J Dev Biol* 51:715–721, 2007.

Morikawa S, Baluk P, Kaidoh T, et al.: Abnormalities in pericytes on blood vessels and endothelial sprouts in tumors, *Am J Pathol* 160:985–1000, 2002.

Nehls V, Denzer K, Drenckhahn D: Pericyte involvement in capillary sprouting during angiogenesis in situ, *Cell Tissue Res* 270:469–474, 1992.

Nicosia RF, Villaschi S: Rat aortic smooth muscle cells become pericytes during angiogenesis in vitro, *Lab Investig* 73:658–666, 1995.

Orlidge A, D'Amore PA: Inhibition of capillary endothelial cell growth by pericytes and smooth muscle cells, *J Cell Biol* 105:1455–1462, 1987.

Ozerdem U, Stallcup WB: Early contribution of pericytes to angiogenic sprouting and tube formation, *Angiogenesis* 6:241–249, 2003.

Ozerdem U, Stallcup WB: Pathological angiogenesis is reduced by targeting pericytes via the NG2 proteoglycan, *Angiogenesis* 7:269–276, 2004.

Patan S: TIE1 and TIE2 receptor tyrosine kinases inversely regulate embryonic angiogenesis by the mechanism of intussusceptive microvascular growth, *Microvasc Res* 56:1–21, 1998.

Patel-Hett S, D'Amore PA: Signal transduction in vasculogenesis and developmental angiogenesis, *Int J Dev Biol* 55:353–363, 2012.

Podestà F, Romeo G, Liu WH, et al.: Bax is increased in the retina of diabetic subjects and is associated with pericyte apoptosis in vivo and in vitro, *Am J Pathol* 156:1025–1032, 2000.

Puri MC, Rossant J, Alitalo K, et al.: The receptor tyrosine kinase TIE is required for integrity and survival of vascular endothelial cells, *EMBO J* 14:5884–5891, 1995.

Rajantie I, Ilmonem M, Alminaite A, et al.: Adult bone marrow-derived cells recruited during angiogenesis comprise precursors for periendothelial vascular mural cells, *Blood* 104:2084–2086, 2004.

Salvucci O, Maric D, Economopolou M, et al.: Ephrin B reverse signaling contributes to endothelial and mural cell assembly into vascular structures, *Blood* 114:1707–1716, 2009.

Sato Y, Rifkin DB: Inhibition of endothelial cell movement by pericytes and smooth muscle cells: activation of a latent transforming growth factor-beta 1-like molecule by plasmin during co-culture, *J Cell Biol* 109:309–315, 1989.

Song S, Ewald AJ, Stallcup W, et al.: PDGFRβ+ perivascular progenitor cells in tumours regulate pericyte differentiation and vascular survival, *Nat Cell Biol* 7:870–879, 2005.
Speiser P, Gittelsohn AM, Patz A: Studies on diabetic rethinopathy. 3. Influence of diabetes on intratumoral pericytes, *Arch Ophthalmol* 80:332–337, 1968.
Stratman AN, Davis GE: Endothelial cell-pericyte interactions stimulate basement membrane matrix assembly: influence on vascular tube remodeling, maturation, and stabilization, *Microsc Microanal* 18:68–80, 2012.
Stratman AN, Malotte KM, Mahan RD, et al.: Pericyte recruitment during vasculogenic tube assembly stimulates endothelial basement membrane matrix formation, *Blood* 114:5091–5101, 2009.
Sundberg C, Kowanetz M, Brown LF, et al.: Stable expression of angiopoietin-1 and other markers by cultured pericytes: phenotypic similarities to a subpopulation of cells in maturing vessels during later stages of angiogenesis in vivo, *Lab Investig* 82:387–401, 2002.
Suri C, Jones PF, Patan S, et al.: Requisite role of angiopoietin-1, a ligand for the Tie-2 receptor, during embryonic angiogenesis, *Cell* 87:1171–1180, 1996.
Thurston G, Suri C, Smith K, et al.: Leakage-resistant blood vessels in mice transgenically overexpressing angiopoietin-1, *Science* 286:2511–2514, 1999.
Tong RT, Boucher Y, Kozin SV, et al.: Vascular normalization by vascular endothelial growth factor receptor 2 blockade induces a pressure gradient across the vasculature and improves drug penetration in tumors, *Cancer Res* 64:3731–3736, 2004.
Uemura A, Ogawa M, Hirashima M, et al.: Recombinant angiopoietin-1 restores higher-order architecture of growing blood vessels in mice in the absence of mural cells, *J Clin Invest* 110:1615–1817, 2002.
Virgintino D, Girolamo F, Errede M, et al.: An intimate interplay between precocius, migrating pericytes and endothelial cells govern human fetal brain angiogenesis, *Angiogenesis* 10:35–45, 2007.
Von Tell D, Armulik A, Betsholtz C: Pericytes and vascular stability, *Exp Cell Res* 312:623–629, 2006.
Wakui S, Yokoo K, Muto T, et al.: Localization of Ang-1, -2, Tie-2, and VEGF expression at endothelial-pericyte interdigitation in rat angiogenesis, *Lab Investig* 86:1172–1184, 2006.
Wang T, Baron M, Trump D: An overview of notch 3 function in vascular smooth muscle cells, *Prog Biophys Mol Biol* 96:499–509, 2007.
Willet CG, Boucher Y, Di Tomaso E, et al.: Direct evidence that the VEGF-specific antibody bevacizumab has antivascular effects in human rectal cancer, *Nat Med* 10:145–147, 2004.
Xu K, Cleaver O: Tubulogenesis during blood vessel formation, *Semin Cell Dev Biol* 22:993–1004, 2011.
Yamagishi S, Yonekura H, Yamamoto Y, et al.: Vascular endothelial growth factor acts as a pericyte mitogen under hypoxic conditions, *Lab Investig* 79:501–509, 1999.
Yonoff M: Diabetic retinopathy, *N Engl J Med* 24:1344–1349, 1966.
Zhang L, Yang N, Park JW, et al.: Tumor-derived vascular endothelial growth factor up-regulates angiopoietin-2 in host endothelium and destabilizes host vasculature, supporting angiogenesis in ovarian cancer, *Cancer Res* 63:3403–3412, 2003.

CHAPTER 14

Arteriogenesis

Functional studies on arteriogenesis began in 1785 with Hunter, who documented his findings on the occlusion of the external carotid artery of a buck and its physiological consequences. He observed that the animal's ipsilateral antler became cool to touch and its arterial pulse became impalpable, as its blood supply was dependent on the ligated artery. However, when the animal was reexamined a week later, the temperature and arterial pulsations of the antler had normalized. On further examination, Hunter found that small branches of the artery above and below the ligature had enlarged, and, through their anastomoses, the blood supply was restored (Murley, 1984).

During embryonic life, differentiation of arteries and veins was believed to be governed by hemodynamic forces, molding these vessels from the primary vascular plexus. Thoma (1893) observed that vessels carrying a lot of blood widen, whereas those that carry little flow regress. Murray (1932) postulated that vessels adapt to flow in order to optimize the shear stress to which they are subjected. These studies have shown that flow can alter lumen dimensions of arterial segments. Cessation of blood flow into a capillary segment causes regression of the vessel, whereas an increase in pressure and shear stress may be an inductive factor for the local recruitment of smooth muscle cells, leading to the differentiation of a capillary vessel into an artery or a vein. The increased flow leads to increased fluid shear stress, which is proportional to the velocity of blood flow. In turn, shear stress acts directly upon gene expression involved in the endothelial cell cycle. Shear stress is critical during embryonic vascular patterning. Defective hierarchical remodeling of the yolk sac vasculature is seen in Mcl1a null embryos with deficient contractility of cardiomyocytes, and this effect can be attributed to shear stress (Lucitti et al., 2007). Vascular endothelial growth factor (VEGF) determines the specification of arterial fate during vasculogenesis, whereas shear stress maintains arterial identity after blood circulation is established. Cell migration in response to hemodynamic cues contributes to artery formation through influencing the number of endothelial cells in enlarging arterial segments (Rochon et al., 2016).

New Insights on the Development of the Vascular System
https://doi.org/10.1016/B978-0-323-90599-2.00014-3

Ligation of the extraembryonic artery induces a morphological and genetic transformation of the arteries into veins and vice versa (Le Noble et al., 2004). Grafts of embryonic quail arteries and veins demonstrated that arterial cells can colonize veins of the host and vice versa (Moyon et al., 2001). Pardanaud et al. (2016) showed that sympathetic innervations promote arterial endothelial cell fate in vivo. Removal of the sympathetic nerves decreases arterial fate and leads to colonization of veins.

Two phases of arteriogenesis can be observed, the proliferating and remodeling phases. Proliferation of the endothelium is followed by smooth muscle cell mitosis, disruption of the lamina elastica interna, migration of vascular smooth muscle cells to form a new neointima, tissue lysis, and cell death of the perivascular tissue to create a space for the growing and expanding new artery. The descending aorta is the first intraembryonic vessel to acquire a tunica media, consisting of smooth muscle cells (Hughes, 1942). Moreover, arterial morphogenesis occurring during embryonic development includes remodeling of the primary embryonic plexus, forming arterial trees. In postnatal life, the term "arteriogenesis" refers to anatomic transformation of preexisting arterioles with increasing lumen area and wall thickness, due to a thick muscular layer and purchasing of viscoelastic and vasomotor capacities (Conway et al., 2001). The growth of arterioles in skeletal muscles results from preexisting capillaries recruiting smooth muscle cells (Hansen-Smith et al., 2001). This term is used to indicate: (i) the extension of an arterial tree occurring during postnatal life (Moraes et al., 2013) and (ii) the transformation of native microvascular collateral arterioles into functional arteries after an ischemic injury with consequent recovery of blood flow (Bloor, 1974; Zhang et al., 2010; Faber et al., 2011; Deindl and Schaper, 2013; Sealock et al., 2014).

Arteriogenesis differs from angiogenesis in several aspects, with the most important being the dependence of angiogenesis on hypoxia and the dependence of arteriogenesis on inflammation. Arteriogenesis occurs in nonhypoxic tissues and has often been used to denote remodeling of the native microvascular collateral arterioles into the arteries. A collateral vessel, resulting from the arteriogenic process, always conducts arterial blood flow and cannot become hypoxic. Collateral vessels in the vascular periphery are surrounded by normoxic tissue. Deindl et al. (2001), by analyzing several time points after occlusion of the femoral artery, could detect neither an increased expression level of hypoxia inducible factor 1 alpha (HIF-1α) mRNA nor an upregulation of the HIF-1-controlled VEGF gene expression. Experimental administration of a constitutively active form of HIF-1α positively influenced arteriogenesis in a hindlimb ischemic model (Patel et al., 2005).

Collateral circulation takes place between the arteries and the arterioles, in healthy tissues and under pathological conditions, through an increase in vessel diameter after an ischemic injury (Faber et al., 2014). In the collateral vessels, connecting two feed arteries or the crowns of the neighboring arterial trees, blood flow occurs in the opposite direction, resulting in a hemodynamic environment of low and oscillatory shear stress and inhibition of hemostatic thrombosis (Trzeciakowski and Chilian, 2008; Chalothorn and Faber, 2010a,b; Toriumi et al., 2009). There are almost two different modalities of formation of collateral arteries: (i) Artery-to-artery anastomosis bypasses the capillary bed to provide blood flow to tissues served by an occluded artery. These connections show minimal tortuosity and do not increase their lumen after an ischemic injury (Hillmeister et al., 2011; Hecht et al., 2012). Numerous connections have been demonstrated between branches of the same and different coronary arteries in human hearts. Moreover, the circle of Willis connects major cerebral arteries at the base of the brain. Finally, the interconnections between the arteria thoracica interna and the arteria epigastrica inferior bypass any obstructions in the descending thoracic aorta down to the iliac artery. (ii) The second modality includes arteriole-to-arteriole anastomosis, interconnecting a small portion of arterioles of the neighboring arterial trees. They are present in most healthy tissues, as pial collaterals of the brain and spinal cord, skeletal muscle, and skin, with an average diameter $< 100\,\mu m$ (Shoja et al., 2010), and in the heath, with an average diameter $> 150\,\mu m$ (Zweifach, 1973), whereas they are absent in other tissues, such as the retina and noncapsular kidney (Shoja et al., 2010). They are tortuous (Tomanek, 2013; Schaper, 2009; Van Royen et al., 2010) and their number, diameter, and remodeling seem to be genetically determined (Zhang et al., 2010; Chalothorn and Faber, 2010a,b). During embryonic development, microvascular collaterals develop after the arterial trees (Chalothorn and Faber, 2010a,b; Lucitti et al., 2012). These collaterals increase their diameter and wall thickness, as a consequence of the growth of a muscular layer, whereas their number decreases. This is completed a few weeks after birth.

Under physiological conditions, as a result of chronic exercise or muscle loading, there is an increase in the number and length of distal arterioles, with extension of an arterial tree (Hansen-Smith et al., 2001). Under pathological conditions, arteriogenesis is associated with a degradation of the basement membrane, a modification of smooth muscle cells from a contractile to a proliferative phenotype associated with loss of desmin (Wolf et al., 1998), and an inflammatory reaction around vessels. As a consequence, the vessels are augmented in diameter and wall thickness (Wolf et al., 1998; Schaper et al., 1976).

Arteriogenesis is the most efficient mechanism to increase blood flow in an ischemic region (Scholz et al., 2001). Femoral artery occlusion leads to ischemia of the lower leg, while collaterals develop in parallel to the stenosis (Van Royen et al., 2010). The initial stage of collateralization is characterized by a wave of monocyte-driven inflammation and vasodilation; in the second phase, the inflammatory response is resolved and the newly formed arteries mature and the wall stabilizes to accommodate the increase in blood conductance. Arteriogenesis eliminates the signs and symptoms of severe vascular occlusion, such as a complete occlusion of a major coronary artery (Schaper, 2009).

References

Bloor CM: Functional significance of the coronary collateral circulation. A review, *Am J Pathol* 76:561–588, 1974.

Chalothorn D, Faber JE: Formation and maturation of the native cerebral collateral circulation, *J Mol Cell Cardiol* 49:251–259, 2010a.

Chalothorn D, Faber JE: Strain-dependent variation in collateral circulatory function in mouse hindlimb, *Physiol Genomics* 42:469–479, 2010b.

Conway EM, Collen D, Carmeliet P: Molecular mechanisms of blood vessel growth, *Cardiovasc Res* 49:507–521, 2001.

Deindl E, Schaper W: Editorial: avenue to arteriogenesis, *Curr Vasc Pharmacol* 11:2–4, 2013.

Deindl E, Buschmann I, Hoefer IE, et al.: Role of ischemia and of hypoxia-inducible genes in arteriogenesis after femoral artery occlusion in the rabbit, *Circ Res* 89:779–786, 2001.

Faber JE, Zhang H, Lassance-Soares RM, et al.: Aging causes collateral rarefaction and increased severity of ischemic injury in multiple tissues, *Arterioscler Thromb Vasc Biol* 31:1748–1756, 2011.

Faber JE, Chilian WM, Deindl E, et al.: A brief etymology of the collateral circulation, *Arterioscler Thromb Vasc Biol* 34:1854–1859, 2014.

Hansen-Smith F, Egginton S, Zhou AL, et al.: Growth of arterioles precedes that of capillaries in stretch-induced angiogenesis in skeletal muscle, *Microvasc Res* 62:1–14, 2001.

Hecht N, He J, Kremenetskaia I, et al.: Cerebral hemodynamic reserve and vascular remodeling in C57/BL6 mice are influenced by age, *Stroke* 43:3052–3062, 2012.

Hillmeister P, Gatzke N, Dülsner A, et al.: Arteriogenesis is modulated by bradykinin receptor signalling, *Circ Res* 109:524–533, 2011.

Hughes AFW: The histogenesis of arteries of the chick embryo, *J Anat* 77:266–287, 1942.

Le Noble F, Moyon D, Pardanaud L, et al.: Flow regulates arterial-venous differentiation in the chick embryo yolk sac, *Development* 131:361–375, 2004.

Lucitti JL, Jones EA, Huang C, et al.: Vascular remodelling of the mouse yolk sac requires hemodynamic force, *Development* 134:3317–3326, 2007.

Lucitti JL, Mackey JK, Morrison JC, et al.: Formation of the collateral circulation is regulated by vascular endothelial growth factor-A and a disintegrin and metalloprotease family members 10 and 1, *Circ Res* 111:1539–1550, 2012.

Moraes F, Paye J, Mac Gabhann F, et al.: Endothelial cell-dependent regulation of arteriogenesis, *Circ Res* 113:1076–1086, 2013.

Moyon D, Pardanaud L, Yuan L, et al.: Plasticity of endothelial cells during arterial-venous differentiation in the avian embryo, *Development* 128:3359–3370, 2001.

Murley R: John Hunter, velvet and vascular surgery, *Ann R Coll Surg Engl* 66:214–218, 1984.

Murray FP: The development in vitro of the blood of the early chicken embryo, *Proc R Soc Lond B* 111:497–521, 1932.

Pardanaud L, Pibouin-Fragner L, Dubrac A, et al.: Sympathetic innervation promotes arterial fate by enhancing endothelial ERK activity, *Circ Res* 119:607–620, 2016.

Patel TH, Kimura H, Weiss CR, et al.: Constitutively HIF-alpha improves perfusion and arterial remodelling in an endovascular model of limb ischemia, *Cardiovasc Res* 68:144–154, 2005.

Rochon ER, Menon PG, Roman BL: Alk1 controls arterial endothelial cell migration in lumenized vessels, *Development* 143:2593–2602, 2016.

Schaper W: Collateral circulation: past and present, *Basic Res Cardiol* 104:5–21, 2009.

Schaper J, König R, Franz D, et al.: The endothelial surface of growing coronary collateral arteries: intimal margination and diapedesis of monocytes: a combined SEM and TEM study, *Virchows Arch A Pathol Anat Histol* 370:193–205, 1976.

Scholz D, Cai WJ, Schaper W: Arteriogenesis, a new concept of vascular adaptation in occlusive disease, *Angiogenesis* 4:247–257, 2001.

Sealock R, Zhang H, Lucitti JL, et al.: Congenic fine-mapping identifies a major causal locus for variation in the native collateral circulation and ischemic injury in brain and lower extremity, *Circ Res* 114:660–671, 2014.

Shoja MM, Tubbs RS, Loukas M, et al.: The sub-peritoneal arterial plexus of sir William turner, *Ann Anat* 192:194–198, 2010.

Thoma R: *Untersuchungen über die histogenese und histomechanik des gefssytems*, Stuttgart, 1893, Ferdinand Enke.

Tomanek RJ: *Coronary vasculature*, New York, 2013, Springer.

Toriumi H, Tatarishvili J, Tomita M, et al.: Dually supplied T-junctions in arteriolo-arteriolar anastomosis in mice: key to local hemodynamic homeostasis in normal and ischemic states? *Stroke* 40:3378–3383, 2009.

Trzeciakowski J, Chilian WM: Chaotic behavior of the coronary circulation, *Med Biol Eng Comput* 46:433–442, 2008.

Van Royen N, Piek JJ, Schaper W, et al.: A critical review of clinical arteriogenesis research, *J Am Coll Cardiol* 55:17–25, 2010.

Wolf C, Cai WJ, Vosschulte R, et al.: Vascular remodeling and altered protein expression during growth of coronary collateral arteries, *J Mol Cell Cardiol* 30:2291–2305, 1998.

Zhang H, Prabhakar P, Sealock R, et al.: Wide genetic variation in the native pial collateral circulation is a major determinant of variation in severity of stroke, *J Cereb Blood Flow Metab* 30:923–934, 2010.

Zweifach BW: Microcirculation, *Annu Rev Physiol* 35:117–150, 1973.

CHAPTER 15

Arterial and venous specification

Blood flows away from the heart via the arteries delivering oxygen and nutrients to peripheral organs and returns to the heart via the veins. Until the late 1990s, the distinction between the arteries and veins was recognized solely based on anatomical and functional differences. A formation of anatomically and functionally distinct arteries and veins is required for the establishment of a functional cardiovascular system.

The dorsal aorta is the first blood vessel to form. Angioblasts begin expressing arterial markers Dll-4 and connexin 37 (cx37) in the endothelial cords of the dorsal aorta and subsequently gain more arterial markers such as ephrin-B2 (Eph-B2) and cx40 during the lumenization process (Chong et al., 2011). Slightly later, other groups of mesodermal progenitors contribute to vasculogenesis of the endocardium, the sinus venosus, the vitelline veins, and the cardinal veins, thus establishing a primordial circulatory system (Chong et al., 2011).

With the exception of capillaries, all blood vessel branches of the vascular tree contain a tunica intima (tunica interna), which is lined with endothelial cells, supported by a basement membrane. The arteries have an inner (or internal) elastic lamina, whereas both the arteries and veins contain a tunica media composed of concentric rings of smooth muscle cells supported by connective tissue, such as collagen fibers. The arteries also have an outer (or external) elastic lamina, whereas both vessels have a tunica externa (tunica adventitia). The endothelial cells of the arteries are elongated and aligned in the direction of blood flow and are surrounded by multiple layers of smooth muscle cells. However, the veins have rounder endothelial cells that do not align with flow, lack elastic laminae, and many possess valves that project into the lumen.

In a developing vertebrate embryo, the dorsal aorta and cardinal vein establish the first closed circulatory loop. Although most of our knowledge regarding the mechanisms of arterial-venous specification has been gleaned from this vascular system, alternate or additional pathways may be implicated in this specification of other embryonic or postnatal vascular beds. Sprouts first develop from the dorsal aorta in response to vascular endothelial

New Insights on the Development of the Vascular System
https://doi.org/10.1016/B978-0-323-90599-2.00015-5

growth factor (VEGF) giving rise to intersegmental vessels, which migrate dorsally and anastomose to form a network of arterial vessels that lacks a venous drainage (Zygmunt et al., 2012). Subsequently, bone morphogenetic protein (BMP) signaling induces sprouting from the cardinal vein and some of these connect to the intersegmental vessels, providing a venous drainage (Wiley et al., 2011). In vivo studies in zebrafish have demonstrated that patterning of intersegmental vessels in the trunk depends on differences in flow in which initiation of arterial flow in an intersegmental vessel leads to a connection of neighboring intersegmental vessels to secondary sprouts to become venous (Isogai et al., 2003).

During vasculogenesis, the initial arterial-venous specification occurs prior to the initiation of blood circulation, suggesting that this specification is genetically programmed and not driven by blood flow although hemodynamic forces play a major role in endothelial cell fate once the flow starts (Garcia-Cardena and Slegtenhorst, 2016).

Differentiation of the arteries and veins was believed to be governed by hemodynamic forces, molding these vessels from the primary vascular plexus. Thoma (1893) observed that vessels that carry a lot of blood widen, whereas those that carry little flow regress. Murray (1926) postulated that vessels adapt to flow in order to optimize the shear stress to which they are subjected. These studies have shown that flow can alter lumen dimensions of arterial segments. In 1952, Stéphan (1952) demonstrated that rerouting flow by artificially obstructing the vitelline artery results in perfusion of the arterial tree with blood of venous origin, which transforms the arteries into veins.

Otherwise, labeling experiments conducted in zebrafish indicate that the arterial and venous fate of endothelial precursors may be determined before the formation of the blood vessels (Zhong et al., 2001). These authors followed individual angioblasts and found that all the progeny of a single angioblast formed either veins or arteries but never both. The discovery that members of the Eph family are differentially expressed in the arteries and veins from extremely early stages of development, prior to development of a functional circulation, was one of the first indications that the artery-vein identity is intrinsically programmed. Ephs are transmembrane ligands that interact with Eph receptors and mediate bidirectional signaling: reverse signaling through the cytoplasmic PDZ domain of Eph and forward signaling into the Eph receptor tyrosine kinase.

Mutations of Eph-B2 and Eph-B4 both lead to early embryonic lethality at around E9.5 (Wang et al., 1998; Adams et al., 1999; Gerety et al.,

1999; Gerety and Anderson, 2002). Remodeling of the primary vascular plexus into the arteries and veins was arrested in both mutants. These findings suggest important roles for Eph-B2/Eph-B4 interactions on arterial and venous endothelial cells, respectively. Eph-B2, which is concentrated on the filopodia of tip cells (Wang et al., 2010), marks arterial endothelial cells and smooth muscle cells, whereas Eph-B4, a receptor for Eph-B2, only marks veins. Mutations in the PDZ domain of Eph-B2 result in reduction of vascular sprouting and filopodia extensions, whereas overexpression of Eph-2 increases filopodia and motility responsible for the alteration of mature vascular formation (Sawamiphak et al., 2010; Wang et al., 2010). When Eph-B2 is compromised, VEGF receptors (VEGFRs) are not endocytosed, thus reducing signaling output (Sawamiphak et al., 2010; Wang et al., 2010). Moreover, Eph-B2–Eph-B4 signaling participates in the formation of arteriovenous anastomoses through arresting VEGF- and angiopoietin-1 (Ang-1)-induced endothelial cell proliferation/migration at the arterial-venous interface (Wang et al., 1998; Gerety et al., 1999; Zhong et al., 2001). Disruption of the Eph-B2 gene in mice results in retarded growth and embryonic lethality at E10.5. Vasculogenesis is halted at the primary plexus stage, endothelial cells are disorganized, and features of angiogenic remodeling are absent (Adams and Klein, 2000).

Other specific markers for the arteries and veins include for the arterial system neuropilin-1 (NRP-1) and members of the Notch family, including Notch-3, DDL4, and GRIDLOCK (Grl), described in zebrafish, chick, and mouse (Herzog et al., 2001; Lawson et al., 2001; Moyon et al., 2001a,b; Shutter et al., 2000; Villa et al., 2001). Venous markers include NRP-2 (Herzog et al., 2001), which at later developmental stages, becomes restricted to lymphatic vessels in chick and mouse (Herzog et al., 2001; Zhong et al., 2001). The different expressions of NRP-1 and NRP-2 in the arteries and veins of early chick and mouse embryos (Herzog et al., 2001; Moyon et al., 2001a,b; Yuan et al., 2002) suggest that NRPs may regulate the segregation of the vascular system into arteries and veins. Herzog et al. (2005) examined the expression patterns of NRP-1 and NRP-2 during the early stages of vasculogenesis and concluded that before the initiation of flow, the primitive vessels of the extraembryonic vascular plexus are already segregated into veins and arteries. The vessels in the venous and arterial parts of the plexus are located around the regions in which the vitelline vein and the vitelline artery will later be formed. Eph-B2 expression cannot be seen in the arterial part of the extraembryonic vascular plexus of 13-somite chick embryos even though the expression of NRP is already segregated, suggesting that Eph-B2, in contrast to NRP-1, is a relatively late marker of the arteries. Loss of Notch signaling in zebrafish embryos leads to an expansion of venous markers such as Eph-B4 and Flt-4 into the arterial domain and overexpression activates arterial markers, such as Eph-B2 into the venous vessels (Lawson et al., 2001). Similar findings have been observed in mammalian vascular development, thus confirming that loss of Notch signaling components causes severe vascular defects (Krebs et al., 2000).

Grl mutants exhibit an aberrant development of the aorta (Weinstein et al., 1995). Notch signaling lies upstream of Grl and is required for arterial-venous differentiation (Zhong et al., 2001; Lawson et al., 2001), and when the Notch signaling pathway activates Grl, it directs endothelial precursors to assume an arterial fate. Endothelial cell migration against the flow direction is associated with artery formation. In zebrafish, vein-derived endothelial tip cells migrate to the emerging arteries during zebrafish caudal regeneration, using CXCL12/CXCR4 signaling (Xu et al., 2014).

Experiments in chick embryos have shown that while blood flow is dispensable for establishment of the arterial-venous identity, it is required for maintaining arterial-venous fate (Moyon et al., 2001a; le Noble et al., 2004), suggesting that a degree of plasticity exists in early arterial and venous identity. Indeed, arterial-venous gene expression patterns can be reset if blood flow is reversed in the chick yolk sac (le Noble et al., 2004).

References

Adams RH, Klein R: Eph receptors and ephrin ligands. Essential mediators of vascular development, *Trends Cardiovasc Med* 10:183–188, 2000.

Adams RH, Wilkinson GA, Weiss C, et al.: Roles of ephrinB ligands and EphB receptors in cardiovascular development: demarcation of arterial/venous domains, vascular morphogenesis, and sprouting angiogenesis, *Genes Dev* 13:295–306, 1999.

Chong DC, Koo Y, Xu K, et al.: Stepwise arteriovenous fate acquisition during mammalian vasculogenesis, *Dev Dyn* 240:2153–2165, 2011.

Garcia-Cardena G, Slegtenhorst BR: Hemodynamic control of endothelial cell fates in development, *Annu Rev Cell Dev Biol* 32:633–648, 2016.

Gerety SS, Anderson DJ: Cardiovascular ephrinB2 is essential for embryonic angiogenesis, *Development* 129:1397–1410, 2002.

Gerety SS, Wang HU, Chen ZF, et al.: Symmerical mutant phenotypes of the receptor EphB4 and its specific transmembrane ligand ephrin-B2 in cardiovascular development, *Mol Cell* 4:403–414, 1999.

Herzog Y, Kalcheim C, Kahane N, et al.: Differential expression of neuropilin-1 and neuropilin-2 in arteries and veins, *Mech Dev* 109:115–119, 2001.

Herzog Y, Guttmann-Raviv N, Neufeld G: Segregation of arterial and venous markers in subpopulations of blood islands before vessel formation, *Dev Dyn* 232:1047–1055, 2005.

Isogai S, Lawson ND, Torrealday S, et al.: Angiogenic network formation in the developing vertebrate trunk, *Development* 130:5281–5290, 2003.

Krebs LT, Xue Y, Norton CR, et al.: Notch signaling is essential for vascular morphogenesis in mice, *Genes Dev* 14:1343–1352, 2000.

Lawson ND, Scheer N, Pham VN, et al.: Notch signaling is required for arterial-venous differentiation during embryonic vascular development, *Development* 128:3675–3683, 2001.

le Noble F, Moyon D, Pardanaud L, et al.: Flow regulates arterial-venous differentiation in the chick embryo yolk sac, *Development* 131:361–375, 2004.

Moyon D, Pardanaud L, Yuan L, et al.: Plasticity of endothelial cells during arterial-venous differentiation in the avian embryo, *Development* 128:3359–3370, 2001a.

Moyon D, Pardanaud L, Yuan L, et al.: Selective expression of angiopoietin 1 and 2 in mesenchymal cells surrounding veins and arteries of the avian embryo, *Mech Dev* 106:133–136, 2001b.

Murray P: The physiological principle of minimum. The vascular system and the cost of blood volume, *Proc Natl Acad Sci U S A* 12:207–214, 1926.

Sawamiphak S, Seidel S, Essmann CL, et al.: Ephrin-B2 regulates VEGFR2 function in developmental and tumour angiogenesis, *Nature* 465:487–491, 2010.

Shutter JR, Scully S, Fan W, et al.: Dll4, a novel notch ligand expressed in arterial endothelium, *Genes Dev* 14:1313–1318, 2000.

Stéphan F: Contribtion expérimentale à l'étude du development du système circulatoire chez l'embryon de poulet, *Bull Biol Fr Belg* 86:218–310, 1952.

Thoma R: *Untersuchungen über die histogenese und histomechanik des gefssytems*, Stuttgart, 1893, Ferdinand Enke.

Villa N, Walker L, Lindsell CE, et al.: Vascular expression of notch pathway receptors and ligands is restricted to arterial vessels, *Mech Dev* 108:161–164, 2001.

Wang H, Chen Z, Anderson D: Molecular distinction and angiogenic interaction between embryonic arteries and veins revealed by ephrin-B2 and its receptor Eph-B4, *Cell* 93:741–753, 1998.

Wang Y, Nakayama M, Pitulescu ME, et al.: Ephrin-B2 controls VEGF-induced angiogenesis and lymphangiogenesis, *Nature* 465:483–486, 2010.

Weinstein BM, Stemple DL, Driever W, et al.: Gridlock, a localized heritable vascular patterning defect in the zebrafish, *Nat Med* 1:1143–1147, 1995.

Wiley DM, Kim JD, Hao J, et al.: Distinct signalling pathways regulate sprouting angiogenesis from the dorsal aorta and the axial vein, *Nat Cell Biol* 13:686–692, 2011.

Xu C, Hasan SS, Schmidt I, et al.: Arteries are formed by vein-derived endothelial tip cells, *Nat Commun* 5:5785, 2014.

Yuan L, Moyon D, Pardanaud L, et al.: Abnormal lymphatic vessel development in neuropilin 2 mutant mice, *Development* 129:4797–4806, 2002.

Zhong TP, Childs S, Leu JP, et al.: Gridlock signaling pathway fashions in the first embryonic artery, *Nature* 414:216–220, 2001.

Zygmunt T, Trzaska S, Edelstein L, et al.: 'In parallel' interconnectivity of the dorsal longitudinal anastomotic vessels requires both VEGF signaling and circulatory flow, *J Cell Sci* 125:5159–5167, 2012.

CHAPTER 16

Role of oxygen

Oxygen tension is a crucial factor in new vessel growth, with regions of hypoxia that induce the expression of angiogenic molecules such as vascular endothelial growth factor-A (VEGF-A), and hypoxia stimulates vasculogenesis (Hoper and Jahn, 1995). Hypoxia-inducible factor-1α (HIF-1α) is a transcription factor that is selectively stabilized and activated under hypoxic conditions and that coordinates the adaptive response of tissues to hypoxia (Semenza, 1999). Hypoxia and the subsequent alteration of HIF-1α, can induce VEGF-A and stromal cell-derived factor-1 (SDF-1), which stimulate migration of endothelial precursors during vasculogenesis (Ramirez-Bergeron et al., 2006).

Hypoxia controls VEGF expression by HIF-1α stabilization in an active complex with the aryl hydrocarbon receptor nuclear translocator (ARNT). ARNT null embryos exhibit defective yolk sac angiogenesis and abnormal development of the vitello-embryonic circulation (Maltepe et al., 1997). Homozygote deletion of the von Hippel-Lindau (VHL) factor that is involved in HIF-1α ubiquitination is lethal and is associated with defects in placental vasculogenesis (Gnarra et al., 1997).

Hypoxia leads to an expression of chemotactic factors inducing migration of angioblasts and endothelial precursor cells (EPCs) toward the region to be vascularized by vasculogenesis.

The loss of HIF-1α results in early embryonic lethality characterized by defective vascularization of the developing embryos and the yolk sac (Iyer et al., 1998). Mice that lack a functional HIF-1α locus die in utero due to defects in vascular remodeling (Kotch et al., 1999). Targeted mutations of HIF-1α in mice lead to a series of malformations of the embryo, including abnormal neural folding, myocardial hyperplasia, and defects in both embryonic and extraembryonic vasculatures (Iyer et al., 1998). VEGF-A expression is elevated in HIF-1$\alpha^{-/-}$ embryos (Kotch et al., 1999), suggesting that factors other than VEGF-A are responsible for the observed defective endothelial cell development. The complete absence of HIF-2α leads to severe vascular defects in the yolk sac and embryo proper, and mutant embryos die

New Insights on the Development of the Vascular System
https://doi.org/10.1016/B978-0-323-90599-2.00016-7

by day 12.5, suggesting that HIF-2α plays a critical role in controlling vascular remodeling during early embryonic development (Peng et al., 2000).

The transcription factor FOX01 activates the genes involved in the regulation of cell polarity and directional migration of tip cells toward hypoxic regions and cooperates with a YAP/TAZ-associated kinase, the mammalian sterile 20-like kinase 1 (NST1) (Kim et al., 2019).

HIF-1α promotes cellular glucose uptake and glycolysis through upregulation ofVEGF-A, glucose transporter type 1 (GLUT1), and glycolytic enzymes, including lactate dehydrogenase A (LDHA) and PFKFB3 (Ebert et al., 1996).

References

Ebert BL, Gleadle JM, O'Rourke JF, et al.: Isoenzyme-specific regulation of genes involved un energy metabolism by hypoxa; sililarities with the regulation of erythropoietin, *Biochem J* 313:808–814, 1996.

Gnarra JR, Ward JM, Porter FD, et al.: Defective placental vasculogenesis causes embryonic lethality in VHL-deficient mice, *Proc Natl Acad Sci U S A* 94:9012–9107, 1997.

Hoper J, Jahn H: Influence of environmental oxygen concentration on growth and vascular density of the area vasculosa in chick embryos, *Int J Microcirc Clin Exp* 15:186–192, 1995.

Iyer N, Kotch V, Agani LE, et al.: Cellular and developmenta control of O_2 homeostasis by hypoxia-inducible factor 1α, *Genes Dev* 12:149–162, 1998.

Kim YH, Choi J, Yang MJ, et al.: MST1-FOX01 cascade establishes endothelial tip polarity and facilitates sprouting angiogenesis, *Nat Commun* 10:838, 2019.

Kotch LE, Iyer NV, Laughner E, et al.: Defective vascularization of HIF-1α-null embryos is not associated with VEGF deficiency but with mesenchymal cell death, *Dev Biol* 209:254–267, 1999.

Maltepe E, Schmidt JV, Baunoch D, et al.: Abnormal angiogenesis and responses to glucose and oxygen deprivation in mice lacking the protein ARNT, *Nature* 386:403–407, 1997.

Peng J, Zhang L, Drysdale L, et al.: The transcription factor EPAS-1/hypoxia inducible factor-2α plays an important role in vascular remodeling, *Proc Natl Acad Sci U S A* 97:8368–8391, 2000.

Ramirez-Bergeron DL, Runge A, Adelman DM, et al.: HIF-dependent hematopoietic factors regulate the development of the embryonic vasculature, *Dev Cell* 11:81–92, 2006.

Semenza GL: Regulation of mammalian O_2 homeostasis in hypoxia-inducible factor-1, *Annu Rev Cell Dev Biol* 15:551–578, 1999.

CHAPTER 17

Epigenetic regulation of vascular development (microRNAs)

The vast majority of cells in the human soma are genetically identical, and the genome of somatic cells is stable across the human lifespan. During evolution of species, most new mutations are either disadvantageous to the organism or neutral, not providing an advantage or a disadvantage, and are therefore not leading to their relative increase in the gene pool. Only a minority of new mutations are propagated in the following generations, by increasing the fitness of the affected organism and its progeny. Heritability is a measure of how well differences in people's genes account for differences in their traits; an estimate of the heritability of a trait is specific to one population in one environment, and it can change over time as circumstances change. There are some conditions in which genetic changes are almost exclusively responsible for causing the condition. These are called genetic disorders or inherited diseases.

The term "epigenetics" was first coined in 1942 by Conrad H. Waddington, a British developmental biologist, embryologist, and geneticist at Cambridge University. When Waddington first used the term, little was known about genes and their hereditary role. Until the 1950s, the term was used to describe the events that helped a fertilized egg to become a mature organism. The term "epigenetic" is used to refer to heritable alterations in chromatin that are not due to changes in DNA sequences (Berger et al., 2009). The molecular foundation of epigenetics comprises three highly interconnected pathways: DNA methylation, histone posttranslational modifications, and RNA-based mechanisms. DNA methylation regulates genomic activity. It consists of the addition of methyl groups to the dinucleotide CpG in the context of the so-called CpG islands by the DNA methyltransferase (Dnmts) family enzymes to create 5-methylcytosine (Jin et al., 2011; Miranda and Jones, 2007). As a consequence, the biophysical properties of the DNA change, impairing the recognition by some proteins while allowing the binding of others with a consequent repression of transcription (Prokhortchouk and Defossez, 2008).

New Insights on the Development of the Vascular System
https://doi.org/10.1016/B978-0-323-90599-2.00017-9

The core histones H2A, H2B, H3, and H4 are the major structural components of chromatin. Each nucleosome comprises two copies of each of the four core histones. Eukaryotic DNA is wrapped around histone octamers and forms nucleosomes that are themselves folded into higher-order chromatin structures. Histones have N-terminal tails extending from the compact nucleosomal cores that affect both histone interaction and gene regulation. Some of the histone modifications are generally associated with active genes, such as histone acetylation, whereas others can be associated with both active and repressed genes, like histone methylation. HAT acetylation increases accessibility and promotes transcription, whereas HDAC dampens histone-DNA and histone-nonhistone protein interactions and opposes the activity of HATs (Sasaki and Matsui, 2008). miRNAs play an important role in regulating gene expression at the level of transcription and posttranscription. They are small 21–25 nucleotide noncoding RNAs, which are present in almost all human tissues, modulate hundreds of genes simultaneously, and control multiple signaling pathways involved in the regulation of apoptosis, cell proliferation, differentiation, and migration (Garzon et al., 2006). More than 2500 human miRNAs have been recorded (Kurozumi et al., 2017). Cellular miRNAs can be released in the extracellular milieu membrane-free package inside microvesicles derived from the plasma membrane or within exosomes (Wang et al., 2010).

Epigenetics is involved in many normal cellular processes. Epigenetic modifications can manifest as commonly as the manner in which cells terminally differentiate to end up as skin cells, liver cells, brain cells, etc. Cells, tissues, and organs differ because they have certain sets of genes that are "turned on" or expressed and other sets that are "turned off" or inhibited. Epigenetic silencing is one method to turn off genes, and it can contribute to differential expression. In addition, epigenetics is important for X-chromosome inactivation in female mammals, which is necessary so that females do not have twice the number of X-chromosome gene products as males (Egger et al., 2004). Epigenetic modifications are usually stably heritable through subsequent cell divisions, resulting in permanent changes in gene expression profiles. However, at critical stages in normal development, cells undergo genome-wide epigenetic reprogramming, erasing preexisting epigenetic marks and establishing a new set of marks. For instance, major epigenetic reprogramming occurs at fertilization prior to zygotic development (Morgan et al., 2005).

Hypermethylation of CpG islands in the regulatory regions of certain genes, global hypomethylation, and selective demethylation are involved in the regulation of transcription during angiogenesis (Patra and Bettuzzi, 2007). DNA hypomethylation of angiogenic genes promotes angiogenesis (Xu et al., 2017).

Inhibition of the activity of HIF by HDACi can block angiogenesis. Overexpression of HDAC1 mediates the reduction of p53 and pVHL expressions, resulting in the overexpression of hypoxia inducible factor-1 alpha (HIF-1α) and vascular endothelial growth factor (VEGF) (Kim et al., 2001). Endothelial morphogenesis is HDAC1-dependent and interstitial flow increases the phosphorylation of HDAC1, its activity, and its export from the nucleus. Furthermore, HDAC1 inhibition decreases endothelial morphogenesis and matrix metalloproteinase 14 (MMP14) expression, suggesting that HDAC1 modulates angiogenesis in response to flow (Bazou et al., 2016).

Fibroblast growth factor-2 (FGF-2) and VEGF-A are downregulated by miR-503, miR-497, and miR-126, among which miR-503 binds 3′-UTRs of both FGF-2 and VEGF-A. Inhibition of miRNAs can contribute to the development of novel antiangiogenic therapies. miRNA expression can be controlled by blood flow, thus involving miRNAs in the regulation of vascular differentiation (Liu et al., 2011).

References

Bazou D, Ng MR, Song JW, et al.: Flow induced HDAC1 phosphorylation and nuclear export in angiogenic sprouting, *Sci Rep* 6:34046, 2016.

Berger SL, Kouzarides T, Shiekhattar R, et al.: An operational definition of epigenetics, *Genes Dev* 23:781–783, 2009.

Egger G, Liang G, Aparicio A, et al.: Epigenetics in human disease and prospects for epigenetic therapy, *Nature* 429:457–463, 2004.

Garzon R, Fabbri M, Cimmino A, et al.: MicroRNA expression and function in cancer, *Trends Mol Med* 12:580–587, 2006.

Jin B, Li Y, Robertson KD: DNA methylation: superior or subordinate in the epigenetic hierarchy? *Genes Cancer* 2:607–617, 2011.

Kim MS, Kwon HJ, Lee YM, et al.: Histone deacetylases induce angiogenesis by negative regulation of tumor suppressor genes, *Nat Med* 7:437–443, 2001.

Kurozumi S, Yamaguchi Y, Kurosumi M, et al.: Recent trends in microRNA research into breast cancer with particular focus on the associations between microRNAs and intrinsic subtypes, *J Hum Genet* 62:15–24, 2017.

Liu D, Krueger J, Le Noble F: The role of blood flow and microRNAs in blood vessel development, *Int J Dev Biol* 55:419–429, 2011.

Miranda TB, Jones PA: DNA methylation: the nuts and bolts of repression, *J Cell Physiol* 213:384–390, 2007.

Morgan HD, Santos F, Green K, et al.: Epigenetic reprogramming in mammals, *Hum Mol Genet* 14(spec 1):R47–R58, 2005.

Patra SK, Bettuzzi S: Epigenetic DNA-methylation regulation ofgenes coding for lipid raft-associated components: a role for raft proteins in cell transformation and cancer progression (review), *Oncol Rep* 17:1279–1290, 2007.

Prokhortchouk E, Defossez PA: The cell biology of DNA methylation in mammals, *Biochim Biophys Acta* 1783:2167–2173, 2008.
Sasaki H, Matsui Y: Epigenetic events in mammalian germcell development: reprogramming and beyond, *Nat Rev Genet* 9:129–140, 2008.
Wang K, Zhang S, Weber J, et al.: Export of microRNAs and microRNA-protective protein by mammalian cells, *Nucleic Acids Res* 38:7248–7259, 2010.
Xu Y, Wang Y, Yan S, et al.: Intracellular adenosine regulates epigenetic programming in endothelial cells to promote angiogenesis, *EMBO Mol Med* 9:1263–1278, 2017.

CHAPTER 18

Angiogenesis and organogenesis

Endothelial cells release in a paracrine manner and express on the cell surface different signaling molecules that can affect the density of developing tissues that are intimately associated with them. It has been speculated that the endothelial cell-tissue interaction may "offer the opportunity to control organ development and growth systematically, rather than individually for each organ" (Lammert et al., 2003). Otherwise, other data are consistent with the assumption that the endothelial cell phenotype differs in space and time, thus providing a foundation for the identification of a specific molecular signature to a given microvascular bed.

Endothelial cell phenotypes are diverse not only among different organs but also within the same organ and blood vessel type, as endothelial cells have to fulfill the highly specific functions and needs of different tissues (Aird, 2011). Endothelial heterogeneity has been shown to be mediated by the microenvironment and epigenetic mechanisms (Aird, 2006, 2011). First, endothelial cells are capable of sensing every little change in its surrounding environment, from the inner surface, the outer surface, or its junctions (Aird, 2006). Some of the environmental stimuli act only over short distances, whereas others require a cell-cell contact. In close proximity, a direct cell-cell contact is used, whereas soluble ligands also permit communication over distances.

In this context, environmental factors, such as substrate topography and oxygen tension, modulate maturation of endothelial cells into specific subtypes. The induction of embryonic angioblasts to differentiate into endothelial cells, organize into a vascular network, and populate the specialized vascular bed of an organ results from a complex genetic program, which is not only sensitive to the composition and structure of the extracellular matrix but also is influenced by the cell-cell contact and by angiogenic and angiostatic growth factors and peptides generated by vascular expansion itself.

Endothelial cells react and respond to stimuli to define their phenotype, and, as the stimuli vary from organ to organ, heterogeneity is then generated (Aird, 2011). This microenvironmental stimuli or "input" can be divided in biomechanical or biochemical stimuli. Oxygenation, growth factors,

New Insights on the Development of the Vascular System
https://doi.org/10.1016/B978-0-323-90599-2.00018-0

cytokines, pH, lipoproteins, chemokines, and hormones are some examples of biochemical inputs, whereas hemodynamic fluctuations such as shear stress or strain are included in biomechanical stimuli (Aird, 2006).

Site-specific epigenetic modification plays an important role in conferring heterogeneity to endothelial cells, as these modifications, which include DNA methylation, histone modification, and chromatin remodeling, are reversible and highly dynamic. Moreover, epigenetic factors have a significant role in endothelial cell lineage determination during embryonic development (Aird, 2006).

Antigens are differentially expressed on endothelial cells in certain organs and tissues (Auerbach, 1992). Organ-associated endothelial antigens have been found in brain (Ghandour et al., 1992) and lung capillary endothelial cells (Kennel et al., 1988). Monoclonal antibodies have been generated against tissue-specific endothelial cell surface antigens involved in lymphocyte homing to the mucosal lymphoid tissue (Streeter et al., 1988a) or the peripheral lymphoid tissue (Streeter et al., 1988b). Some of the endothelial markers are constitutively expressed (Table 1), whereas others are expressed only after activation by inflammatory cytokines or growth factors (Table 2) or in specialized endothelial cells (Table 3). The von Willebrand factor (vWF, factor VIII-related antigen) is a multimeric glycoprotein produced by endothelial cells, which can be secreted outside the cell or stored in Weibel-Palade bodies where it serves as a cytoplasmic endothelial marker. Each monomeric unit of vWF has binding sites for factor VIII, platelet glycoprotein Ib (GPIb), GP IIb/IIIa, heparin, and collagen (Meyer et al., 1991). vWF is also produced by megakaryocytes and afterward stored in platelet α-granules, but their release from megakaryocytes remains elusive (Bowie et al., 1986). vWF is considered as a universal marker of endothelial cells, even if it is not expressed uniformly on endothelial cells (Kumar

Table 1 Endothelial markers constitutively expressed.

Markers	Ligands
Factor VIII-related antigen	Factor VIII, platelet glycoprotein Ib, and IIb/IIIa (GPIb, GP IIb/IIIa)
CD31/PECAM-1	CD31
Angiotensin-converting enzyme (ACE/CD143)	Angiotensin
CD34	L-selectin
Endoglin (CD105)	TGF-β1 and -β3 in association with TGF-β receptor type II

Table 2 Inducible endothelial markers.

Markers	Ligands
ICAM-1 (CD54)	LFA-1 integrin
NCAM-1 (CD106)	VLA-4 integrin
E-selectin (CD62E)	Sialyl-Lewis X antigen and other carbohydrates
P-selectin (CD62P)	Carbohydrate determinants on selectin ligands, e.g., PSGL-1
Flt-1 (VEGFR-1)	VEGF
KDR/Flk-1 (VEGFR-2)	
Tie-2/Tek	Angiopoietins

Table 3 Markers of specialized endothelium.

Markers	Preferential expression
GlyCAM-1	HEVs, mammary epithelium
MAdCAM-1	Peyer's patches and mesenteric lymph-node HEVs, venules in intestinal lamina propria, spleen sinus
CD34/Sgp90	Specific glycosylated HEV-restricted CD34 form
VAP-1	HEVs, ECs in large vessels, dendritic cells, smooth muscle cells
VAP-2	HEVs, venules, epithelia, leukocytes
MECA 79	HEVs, chronically inflamed tissue ECs
Flt-4 (VEGFR-3)	Lymphatic ECs
Desmoplakin	Lymphatic ECs, HUVECs, epithelial cells
LuECAM	Lung venous ECs
P-glycoprotein (*mdr* 1a)	BBB
Glut-1	BBB
CD73/transferrin receptor	BBB, erythrocyte precursors, hepatocytes, Kupffer cells
Platelet-derived growth factor receptor	BBB and other cells
HT7	BBB, kidney epithelium, erythroblasts

HEVs, high endothelial venules; *ECs*, endothelial cells; *HUVECs*, human umbilical vein endothelial cells; *BBB*, blood-brain barrier.

et al., 1987; Turner et al., 1987; Smith et al., 1996). It is expressed at higher levels on the venous than on the arterial side of the capillary circulation, in the endothelium of larger vessels and in the adult endocardium in human tissues, and is largely absent from sinusoidal endothelial cells. These differences are maintained when endothelial cells are grown in culture (Smith et al., 1996). Inhibition of vWF expression by siRNAs in endothelial cells caused increased in vitro angiogenesis and increased vascular endothelial growth factor receptor-2 (VEGFR-2)-dependent proliferation and migration coupled with decreased integrin $\alpha v\beta 3$ levels and increased angiopoietin-2 (Ang-2) release; increased vascularization in VWF-deficient mice was demonstrated (Starke et al., 2011). CD31 or platelet endothelial cell adhesion molecule (PECAM-1) is a transmembrane glycoprotein, a member of the immunoglobulin superfamily, and is expressed on early and mature vascular endothelial cells (Müller et al., 2002). Anti-CD31 is the most sensitive pan-endothelial marker available, which recognizes a great number of microvessels. Location of CD31 on the border of two adjacent endothelial cells makes this adhesion molecule a part of the endothelial intercellular junction, which enhances the endothelial barrier properties. The specificity of the antibody to anti-CD31 is not absolute because it reacts with other cells, including platelets (Metzelaar et al., 1991), neutrophils (Watt et al., 1995), T, B, and natural killer cells (Sakurai et al., 2014), hematopoietic progenitor cells (Voermans et al., 2000), and megakaryocytes (Calapso et al., 1992). CD34 is a transmembrane phosphoglycoprotein described as a hematopoietic stem cell (HSC) marker expressed on both myeloid and lymphoid lineages and therefore used as a selective hematopoietic marker (Civin et al., 1984). CD34 is expressed on early and mature vascular endothelial cells and therefore serves as a powerful pan-endothelial marker. This molecule is not expressed on lymphatic endothelial cells under normal conditions and therefore can be used to discriminate between the blood and lymphatic vasculatures. Human endothelial cells display an RGD-containing glycoprotein of 180 kDa called endoglin (CD105) on the plasma membrane (Gougos and Letarte, 1990). Endoglin is a receptor for the transforming growth factor beta (TGF-β) family of proteins that is essential for angiogenesis (Lastres et al., 1996). Endoglin knockdown ($Eng^{-/-}$) mouse embryos die because of vascular and cardiac abnormalities (Mahmoud et al., 2011), and endothelial cells from $Eng^{+/-}$ mice show reduced proliferation and migration and impaired blood vessel formation in vitro and in vivo (Jerkic et al., 2006). Angiotensin-converting enzyme (ACE/CD143) catalyzes the release of histidyl leucine from the –COOH terminus of angiotensin I (AI) to yield the potent vasoactive octapeptide, namely, angiotensin II. The ectoenzyme is primarily located on the luminal membrane of vascular endothelium cells where it hydrolyzes the circulating hormones (Ryan et al., 1976). Endothelial cell activation in response to inflammatory cytokine tumor necrosis factor alpha (TNFα) decreases ACE expression (Saijonmaa et al., 2001). Moreover, ACE is a marker of primitive hematopoietic cells at all stages in the ontogeny of the human hematopoietic system (Jokubaitis et al., 2008).

Liver

In the hepatic primordium, blood vessels provide fundamental morphogenic instructions to hepatocyte precursors (Matsumoto et al., 2001; Yim et al., 2006). Hepatic vessels grow through a combination of angiogenesis and vasculogenesis (Gouysse et al., 2002; Perez-Pomares et al., 2004). In the mouse embryo, prior to liver bud emergence, $CD31^+$ endothelial cell progenitors in the septum transversum closely surround the newly specified foregut hepatic endoderm (Matsumoto et al., 2001).

Close structural interactions occur between the hepatic endoderm and the developing blood vessels early, prior to liver bud emergence. Blood vessels promote hepatic morphogenesis and delimit the mesenchymal domain into which the liver bud grows. Indeed, VEGFR-2 homozygous mutant embryos, which lack endothelial cells but still maintain hepatic specification of the foregut endoderm, express defective liver bud outgrowth (Matsumoto et al., 2001). In the absence of endothelial cells, liver epithelial cells fail to migrate to the surrounding mesenchyme.

In vitro experiments indicate that the critical role of early endothelial cells in the process of liver bud induction occurs prior to the formation of a functioning local vasculature and requires close endothelial/endoderm interactions, involving bone morphogenetic protein-4 (BMP-4) and fibroblast growth factor-8 (FGF-8) (Matsumoto et al., 2001).

Liver sinusoidal endothelial cells (LSECs) are characterized by fenestrae, which do not contain any diaphragm, and the space between LSECs and hepatocytes, the Disse space, is not filled with any basement membrane. Hepatocyte growth factor (HGF) and Wnt2 secreted by LSECs stimulate hepatocyte proliferation (Ding et al., 2010; Le Couter et al., 2001). Inducible endothelial cell-specific deletion of the Wnt secretion mediator Wntless (Wis) and inducible global deletion of the Wnt signaling enhancer Rspo3 impair metabolic liver zonation and liver regeneration due to perturbed Wnt signaling from hepatic endothelial cells to hepatocytes and are characterized by pericentral Gs1 expression (Rocha et al., 2015). The proliferation of axin-2-positive pericentral hepatocytes, a self-renewing reservoir of cells that replace hepatocytes, is controlled by Wnt ligands produced by central vein endothelial cells (Wang et al., 2015). Bone morphogenetic protein-2 (BMP-2) secreted by LSECs acts on hepatocytes to control local and systemic iron metabolism (Koch et al., 2017). The hepatic hormone hepcidin, which is secreted into the bloodstream by hepatocytes, depends on stimulation by BMP-2 secreted by LSECs, and deletion of BMP-2 in LSECs causes a phenotype resembling hemochromatosis (Koch et al., 2017). The deletion of transcription factor Gata-4 in LSECs provokes the transformation of discontinuous sinusoids into continuous capillaries (Fig. 1). The switch from discontinuous LSECs to continuous ones impaired the immigration of hematopoietic progenitor cells into the liver, resulting in anemia and embryonic lethality (Géraud et al., 2017; Olsavszky et al., 2017). HPSC-derived

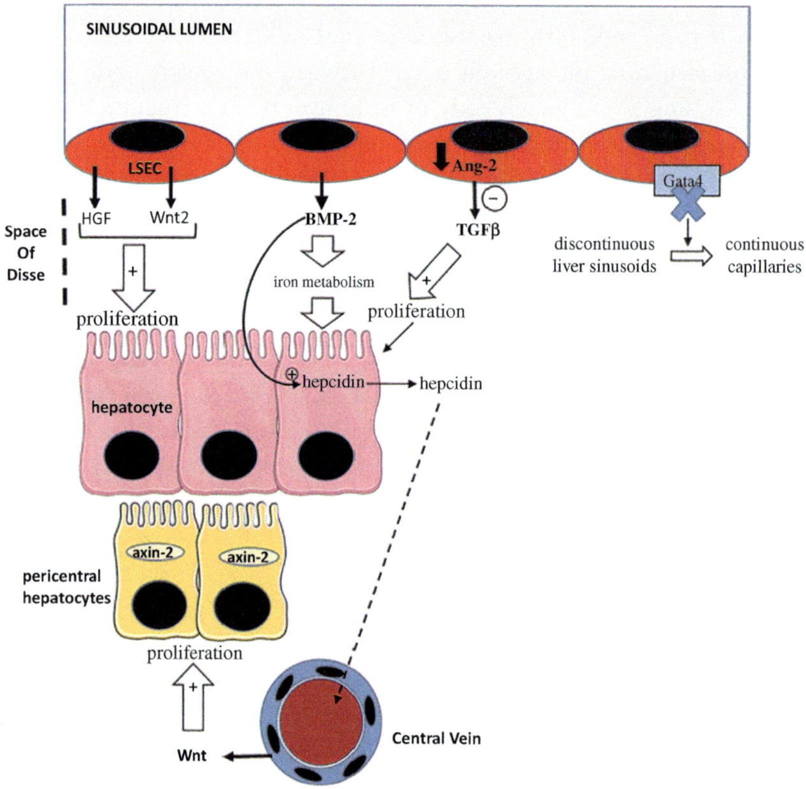

Fig. 1 Complex interplay between liver sinusoid endothelial cells (LSECs) and hepatocytes. Hepatocyte growth factor (HGF) and Wnt2 secreted by liver sinusoidal endothelial cells (LSECs) stimulate hepatocyte proliferation. Proliferation of axin-2-positive pericentral hepatocytes is controlled by Wnt ligands produced by central vein endothelial cells. Bone morphogenetic protein-2 (BMP-2) secreted by LSECs acts on hepatocytes to control local and systemic iron metabolism and induces the secretion of hepcidin into the bloodstream. Downregulation of angiopoietin-2 (Ang-2) in LSECs leads to lower expression of transforming growth factor beta (TGF-β), a negative regulator of hepatocyte proliferation. LSEC-restricted deletion of Gata-4 causes transformation of discontinuous liver sinusoids into continuous capillaries. (Reproduced from Ribatti D, Tamma R, Annese T: The role of vascular niche and endothelial cells in organogenesis and regeneration, *Exp Cell Res* 398:112398, 2021.)

endothelial cells improve the maturation of hepatocytes and engraftment of HPSC hepatocytes (Ramanathan et al., 2015).

Pancreas

In pancreatic development, endothelial cells provide inductive instructions that are critical for the initial differentiation of dorsal pancreatic progenitors, characterized by the expression of the homeobox gene *Pdx1*. Pancreatic cell precursors delaminate from the budding endodermal epithelium at sites where the dorsal pancreatic endoderm is in contact with the endothelium of the aorta (Lammert et al., 2001). Pancreatic beta cells secrete VEGF-A to attract endothelial cells, which form capillaries surrounded by a basement membrane. Laminin-411, a component of the basement membrane, interacts with α6β1 integrin on beta cells to promote insulin gene expression and beta-cell proliferation (Fig. 2) (Nikolova et al., 2007). In VEGFR-2 null mutant embryos, which lack endothelial cells and a vasculature, the dorsal pancreatic bud does not develop (Yoshitomi and Zaret, 2004). In contrast,

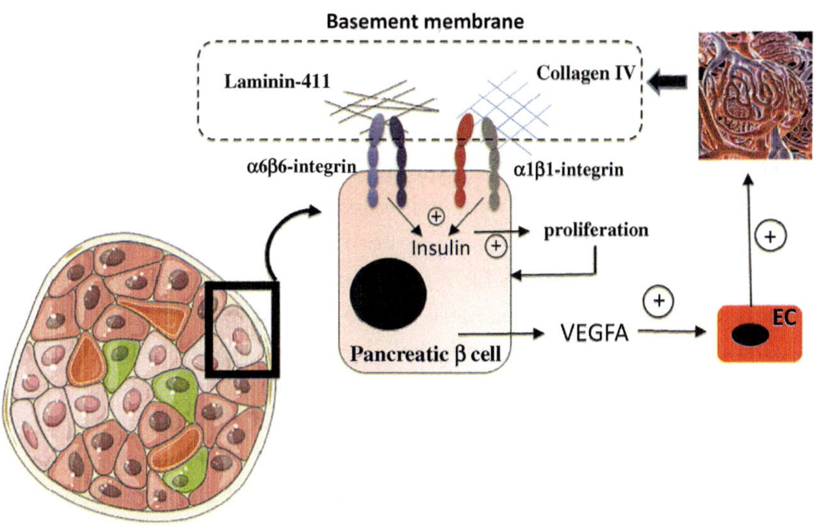

Fig. 2 Pancreatic beta cells secrete VEGF-A, which attracts endothelial cells (ECs) involved in the new formation of capillaries surrounded by a basement membrane. In turn, laminin-411 and collagen IV, two components of the basement membrane, interact with α6β1 integrin and α1β1 integrin, respectively, expressed by pancreatic beta cells, inducing insulin gene expression and beta-cell proliferation. (Reproduced from Ribatti D, Tamma R, Annese T: The role of vascular niche and endothelial cells in organogenesis and regeneration, *Exp Cell Res* 398:112398, 2021.)

the ventral bud proliferates despite a lack of the local vasculature. Cultures of the mouse endoderm initiate insulin expression only when grown in combination with the dorsal aortae but not when grown isolated (Lammert et al., 2001).

In *Xenopus* mutant embryos, the absence of the dorsal aorta leads to strong reduction or lack of marker expressions of development of the dorsal pancreatic anlage, such as *insulin*, *NeuroD*, and *Pax6* (Lammert et al., 2001). Aorta endothelial cells support survival of mesenchyme cells, which, in turn, stimulate initial differentiation of the dorsal pancreatic endoderm to develop the dorsal pancreas (Jacquemin et al., 2006). Coculture of human embryonic stem cell (HESC)-derived pancreatic progenitors (PPs) with human umbilical vein endothelial cells (HUVECs) and rat heart microvascular endothelial cells (RHMVECs) improves PP maturation (Jaramillo et al., 2015). Candiello et al. (2018) generated self-organized organoids from human pluripotent stem cells (HPSCs) and HUVECs grown on hydrogel.

Endocrine gland VEGF (EG-VEGF) is confined to the endocrine organs in human tissues and is able to stimulate proliferation and migration in endothelial cells isolated from the adrenal glands and the pancreas but not from other neuroendocrine organs (Le Couter et al., 2001).

Kidney

In addition to the classically recognized inductive interaction between the metanephritic mesenchyme and the ureteric bud, the interaction between angioblasts and the mesenchyme is involved in kidney development.

Renal blood vessels originate by both vasculogenesis and angiogenesis (Abrahamson et al., 1998). VEGF is responsible of vasculogenesis, glomerulogenesis, and renal tubulogenesis (Tufro et al., 1999). The zebrafish mutant cloche fails to develop endothelia cells and pronephritic glomerulus (Majumdar and Drummond, 1999). Failure of glomerulogenesis is recognized in neonatal mice treated with a soluble VEGFR chimeric protein (Gerber et al., 1999), and mice with ablation of platelet-derived growth factor B (PDGFB) in endothelial cells show impaired glomerulogenesis due to mesangial cell defects (Bjarnegård et al., 2004). Podocytes recruit endothelial cells by expressing VEGF, and, in turn, endothelial cells secrete MMP-2 and PDGF (Bjarnegård et al., 2004).

The ability of vessels to influence the surrounding nonvascular cells may depend on the intrinsic heterogeneity of endothelial cells. In the liver, for instance, the VEGF-A-driven paracrine supply of HGF by sinusoid endothelial cells represents a model for hepatocyte rescue and survival after toxic events. Indeed, HGF induction by VEGF-A or VEGFR-1 agonists may form the basis of a therapeutic scheme aimed toward protection of the liver parenchyma in at least some liver disorders.

The formation of a normal glomerulus depends on the local secretion of matrix metalloproteinase-2 (MMP-2) from endothelial cells (Serluca et al., 2002). MMP-2 enables the assembly of podocytes by degrading extracellular matrix components. Semaphorins 3a and 3c released by podocytes regulate kidney vascular development (Reidy and Tufro, 2011). Both endothelial cells and podocytes synthesize glomerular basement membrane components (Miner, 2012). Endothelial cells regulate the expression of renal genes in progenitor cells, improve glomerular barrier function, and, only in the presence of endothelial cells, the glomeruli are recellularized (Du et al., 2016).

Bone

Vasculature-driven signals turn out to be essential for correct bone morphogenesis (Gerber and Ferrara, 2000). Blood vessel invasion of the cartilage is the critical first step in the cascade of events leading to bone formation. During endochondral bone formation, hypertrophic chondrocytes secrete angiogenic molecules, including transferrin, FGF-2, MMP-9, VEGF, and attract blood vessels from the subchondral bone (Gerber and Ferrara, 2000). These, in turn, promote correct bone formation. VEGF-A coordinates the process of vascular invasion of the epiphyseal growth plate, remodeling of the hypertrophic cartilage, and ossification of the newly formed bone matrix by osteoblasts (Gerber et al., 1999). Indeed, an intravenous injection of soluble VEGFR proteins almost completely abolishes blood vessel development in the epiphyseal growth plate with subsequent impaired trabecular bone formation and ossification. These effects are all completely reverted by withdrawal of anti-VEGF-A treatment.

Endothelial cells participate in osteoinduction through the release of growth factors, by controlling the principal transcription factors required for bone cell differentiation and by facilitating a mineralized tissue within a prevascular network (Grellier et al., 2009). Endothelial cells secrete VEGF, bone morphogenetic proteinn-2 (BMP-2), endothelin-1 (ET-1), and insulin-like growth factors, which affect the migration and proliferation of osteoblasts and the differentiation of osteoprogenitor cells (Bouletreau et al., 2002; Fiedler et al., 2006).

Lung

Lung vascular development occurs by both vasculogenesis and angiogenesis (De Mello et al., 1997). Developing lung vessels come into increasing proximity to epithelial cells, which leads to formation of the functional exchange regions of the alveolus (Burri, 1994). The lungs of neonatal mice treated with a soluble VEGFR chimeric protein are immature with less complex alveolar patterning (Gerber et al., 1999). VEGF-A is secreted by respiratory epithelial cells and creates a VEGF gradient that guides the vascular network to follow and surround the growing bronchi (De Mello et al., 1997; Ng et al., 2001). Neutralizing anti-VEGFR-2 antibodies impair fetal lung maturation (Compernolle et al., 2002). Moreover, loss of VEGFR-2 function reduces lung epithelial cell proliferation and branching (Del Moral et al., 2006). Overexpression of VEGF is responsible for altered lung vascularization and acinar structure (Zeng et al., 1998) as well as increased lung epithelial cell proliferation and branching (Del Moral et al., 2006).

Brain

The cerebral endothelium plays an active role during embryonic neurogenesis (Karakatsani et al., 2019), favoring the differentiation of progenitors into neurons (Lange et al., 2016) and promoting neuron migration through the synthesis of extracellular matrix (Segarra et al., 2018). VEGF mRNAs in the choroid plexus epithelium and periventricular areas induce the ingrowth of capillaries from the perineural vascular plexus (Breier et al., 1992).

Endothelial cells produce artemin and neurotrophin 3 that guide axon extension along vessels of the sympathetic chain ganglia neurons (Carmeliet and Tessier-Lavigne, 2005). Exposure of rat embryos to the angiogenesis inhibitor thalidomide results in vascular malformations and abnormal neuronal development (Hallene et al., 2006). Coculture experiments demonstrated that endothelial cells affected neural stem cell cultures, increasing their self-renewal through a Notch and Hes1 pathway (Shen et al., 2004).

Explants of the dorsal aorta promote adrenergic development in neural crest cultures (Shah et al., 1996; Reissmann et al., 1996). A conditioned medium from isolated cerebral capillary endothelial cells stimulates astrocyte proliferation (Estrada et al., 1990). Endothelial cells express leukemia inhibitory factor (LIF), which is implicated in astrocyte differentiation (Mi et al., 2001). Soluble factors secreted by neural stem cells induce VEGF and brain-derived neurotrophic factor (BDNF) in endothelial cells, resulting in vascular tube formation (Li et al., 2006), and differentiation of neural stem cells into neuroblasts is influenced through endothelial cell BDNF (Leventhal et al., 1999). Deposition of BDNF and SDF1 by endothelial cells in the mouse brain induces proliferation and migration of brain cells to the olfactory bulb (Whitman et al., 2009).

References

Abrahamson DR, Robert B, Hyink DP, et al.: Origins and formation of microvasculature in the developing kidney, *Kidney Int* 54:S7–S11, 1998.

Aird WC: Mechanisms of endothelial cell heterogeneity in health and disease, *Circ Res* 98:159–162, 2006.

Aird WC: Endothelial cell heterogeneity, *Cold Spring Harb Perspect Med* 2, 2011, a006429.

Auerbach R: Endothelial cell heterogeneity, In Simionescu N, Simionescu M, editors: *Endothelial cell dysfunctions*, Boston, 1992, Springer, pp 427–437.

Bjarnegård M, Enge M, Norlin J, et al.: Endothelium-specific ablation of PDGFB leads to pericyte loss and glomerular, cardiac and placental abnorlaities, *Development* 131:1847–1857, 2004.

Bouletreau PJ, Warren SM, Spector JA, et al.: Hypoxia and VEGF up-regulate BMP-2 mRNA and protein expression in microvascular endothelial cells: implications for fracture healing, *Plast Reconstr Surg* 109:2384–2397, 2002.

Bowie EJ, Solberg LA, Fass DN, et al.: Transplantation of normal bone marrow into a pig with severe von Willebrand's disease, *J Clin Invest* 78:26–30, 1986.

Breier G, Albrecht U, Sterrer S, et al.: Expression of vascular endothelial growth factor during embryonic angiogenesis and endothelial cell differentiation, *Development* 114:521–532, 1992.

Burri PH: Fetal and postnatal development of the lung, *Annu Rev Physiol* 46:617–628, 1994.

Calapso P, Vitarelli E, Crisafulli C, Tuccari G: Immunocytochemical detection of megakaryocytes by endothelial markers: a comparative study, *Pathologica* 84:215–223, 1992.

Candiello J, Grandhi TSP, Goh SK, et al.: 3D heterogeneous islet organoid generation from human embryonic stem cells using a novel engineered hydrogel platform, *Biomaterials* 177:27–39, 2018.

Carmeliet P, Tessier-Lavigne M: Common mechanisms of nerve and blood vessel wiring, *Nature* 436:193–200, 2005.

Civin CI, Strauss LC, Brovall C, et al.: Antigenic analysis of hematopoiesis. III. A hematopoietic progenitor cell surface antigen defined by a monoclonal antibody raised against KG-1a cells, *J Immunol* 133:157–165, 1984.

Compernolle V, Brusselmans K, Acker T, et al.: Loss of HIF-2 alpha and inhibition of VEGF impair fetal lung maturation, whereas treatment with VEGF prevents fetal respiratory distress in premature mice, *Nat Med* 8:702–710, 2002.

De Mello DE, Sawyer D, Galvin N, et al.: Early fetal development of of lung vasculature, *Am J Respir Cell Mol Biol* 16:568–581, 1997.

Del Moral PM, Sala FG, Tefft D, et al.: VEGF-A signalling through Flk-1 is a critical facilitator of early embryonic lung epithelial to endothelial crosstalk and branching morphogenesis, *Dev Biol* 290:177–188, 2006.

Ding BS, Nolan DJ, Butler JM, et al.: Inductive angiocrine signals from sinusoidal endothelium are required for liver regeneration, *Nature* 468:310–315, 2010.

Du C, Narayanan K, Leong MF, et al.: Functional kidney bioengineering with pluripotent stem-cell-derived renal progenitor cells and decellularized kidney scaffolds, *Adv Healthc Mater* 5:2080–2091, 2016.

Estrada C, Bready JV, Berliner JA, et al.: Astrocyte growth stimulation by a soluble factor produced by cerebral endothelial cells in vitro, *J Neuropathol Exp Neurol* 49:539–549, 1990.

Fiedler J, Brill C, Blum WF, et al.: IGF-I and IGF-II stimulate direct cell migration of bone-marrow-derived human mesenchymal progenitor cells, *Biochem Biophys Res Commun* 345:1177–1183, 2006.

Géraud C, Koch PS, Zierow J, et al.: GATA4-dependent organ-specific endothelial differentiation controls liver development and embryonic hematopoiesis, *J Clin Invest* 127:1099–1114, 2017.

Gerber HP, Ferrara N: Angiogenesis and bone growth, *Trends Cardiovasc Med* 10:223–228, 2000.

Gerber HP, Vu TH, Ryan AM, et al.: VEGF couple hypertrophic cartilage remodeling, ossification and angiogenesis during endochondral bone formation, *Nat Med* 5:623–628, 1999.

Ghandour S, Langley K, Gombos G, et al.: A surface marker for murine vascular endothelial cells defined by monoclonal antibody, *J Histochem Cytochem* 30:165–170, 1992.

Gougos A, Letarte M: Primary structure of endoglin, an RGD-containing glycoprotein of human endothelial cells, *J Biol Chem* 265:8361–8364, 1990.

Gouysse G, Couvelard A, Frachon S, et al.: Relationship between vascular development and vascular differentiation during liver organogenesis in humans, *J Hepatol* 37:730–740, 2002.

Grellier M, Bordenave L, Amédée J: Cell-to-cell communication between osteogenic and endothelial lineages: implications for tissue engineering, *Trends Biotechnol* 27:562–571, 2009.

Hallene KL, Oby E, Lee BJ, et al.: Prenatal exposure to thalidomide, altered vasculogenesis, and CNS malformations, *Neuroscience* 142:267–283, 2006.

Jacquemin P, Yoshitomi H, Kashima Y, et al.: An endothelial-mesenchymal relay pathway regulates early phases of pancreas development, *Dev Biol* 290:189–199, 2006.

Jaramillo M, Mathew S, Mamiya H, et al.: Endothelial cells mediate islet-specific maturation of human embryonic stem cell-derived pancreatic progenitor cells, *Tissue Eng A* 21:14–25, 2015.

Jerkic M, Rodriguez-Barbero A, Prieto M, et al.: Reduced angiogenic responses in adult endoglin heterozygous mice, *Cardiovasc Res* 69:845–854, 2006.

Jokubaitis VJ, Sinka L, Driessen R, et al.: Angiotensin-converting enzyme (CD143) marks hematopoietic stem cells in human embryonic, fetal, and adult hematopoietic tissues, *Blood* 111:4055–4063, 2008.

Karakatsani A, Shah B, Ruiz de Almodovar C: Blood vessels as regulators of neural stem cell properties, *Front Mol Neurosci* 12:85, 2019.

Kennel SJ, Lankford TK, Ullrich RL, et al.: Enhancement of lung tumor colony formation by treatment of mice with monoclonal antibodies to pulmonary capillary endothelial cells, *Cancer Res* 48:4964–4968, 1988.

Koch PS, Olsavszky V, Ulbrich F, et al.: Angiocrine Bmp2 signaling in murine liver controls normal iron homeostasis, *Blood* 129:415–419, 2017.

Kumar S, West DC, Ager A: Heterogeneity in endothelial cells from large vessels and microvessels, *Differentiation* 36:57–70, 1987.

Lammert E, Cleaver O, Melton D: Induction of pancreatic differentiationby signals from blood vessels, *Science* 294:564–567, 2001.

Lammert E, Cleaver O, Melton D: Role of endothelial cells in early pancreas and liver development, *Mech Dev* 120:59–64, 2003.

Lange C, Turrero Garcia M, Decimo I, et al.: Relief of hypoxia by angiogenesis promotes neural stem cell differentiation by targeting glycolysis, *EMBO J* 35:924–941, 2016.

Lastres P, Letamendia A, Zhang H, et al.: Endoglin modulates cellular responses to TGF-beta 1, *J Cell Biol* 133:1109–1121, 1996.

Le Couter J, Moritz DR, Li B, et al.: Identification of an angiogenic mitogen selective for endocrine gland epithelium, *Nature* 412:877–884, 2001.

Leventhal C, Rafii S, Rafii D, et al.: Endothelial trophic support of neuronal production and recruitment from the adult mammalian subependyma, *Mol Cell Neurosci* 13:450–464, 1999.

Li Q, Ford MC, Lavik EB, et al.: Modeling the neurovascular niche: VEGF- and BDNF-mediated cross-talk between neural stem cells and endothelial cells: an in vitro study, *J Neurosci Res* 84:1656–1668, 2006.

Mahmoud M, Upton PD, Arthur HM: Angiogenesis regulation by TGFbeta signalling: clues from an inherited vascular disease, *Biochem Soc Trans* 39:1659–1666, 2011.

Majumdar A, Drummond LA: Podocyte differentiation in the absence of endothelial cells as revealed in the zebrafish avascular mutant cloche, *Dev Genet* 24:220–229, 1999.

Matsumoto K, Yoshitomi H, Rossant J, et al.: Liver organogenesis promoted by endothelial cells prior to vascular function, *Science* 294:559–563, 2001.

Metzelaar M, Korteweg J, Sixma J, et al.: Biochemical characterization of PECAM-1 (CD31 antigen) on human platelets, *Thromb Haemost* 66:700–707, 1991.

Meyer D, Piétu G, Fressinaud F, et al.: Factor: structure and function, *Mayo Clin Proc* 66:516–523, 1991.

Mi H, Haeberle H, Barres BA: Induction of astrocyte differentiation by endothelial cells, *J Neurosci* 21:1538–1547, 2001.

Miner JH: The glomerular basement membrane, *Exp Cell Res* 318:973–978, 2012.

Müller AM, Hermanns MI, Skrzynski C, et al.: Expression of the endothelial markers PECAM-1, vWF, and CD34 in vivo and in vitro, *Exp Mol Pathol* 72:221–229, 2002.

Ng YS, Sunday ME, Demello DE, et al.: Differential expression ofVEGF isoforms in mouse during development and in the adult, *Dev Dyn* 220:112–121, 2001.

Nikolova G, Strilic B, Lammert E: The vascular niche abd its basement membrane, *Trends Cell Biol* 17:19–25, 2007.

Olsavszky V, Ulbrich F, Singh S, et al.: GATA4 and LMO3 balance angiocrine signaling and autocrine inflammatory activation by BMP2 in liver sinusoidal endothelial cells, *Gene* 627:491–499, 2017.

Perez-Pomares JM, Carmona R, González-Iriarte M, et al.: Contribution of mesothelium-derived cells to liver sinusoids in avian embryos, *Dev Dyn* 229:465–474, 2004.

Ramanathan R, Pettinato G, Beeston JT, et al.: Transplantation of human stem cell-derived hepatocytes in an animal model of acute liver failure, *Surgery* 158:349–359, 2015.

Reidy K, Tufro A: Semaphorins in kidney development and disease: modulators of ureteric bud branching, vascular morphogenesis, and podocyte-endothelial crosstalk, *Pediatr Nephrol* 26:1407–1412, 2011.

Reissmann E, Ernsberger U, Francis-West PH, et al.: Involvement of bone morphogenetic protein-4 and bone morphogenetic protein-7 in the differentiation of the adrenergic phenotype in developing sympathetic neurons, *Development* 122:2079–2088, 1996.

Rocha AS, Vidal V, Mertz M, et al.: The angiocrine factor Rspondin3 Is a key determinant of liver zonation, *Cell Rep* 13:1757–1764, 2015.

Ryan US, Ryan JW, Whitaker C, et al.: Localization of angiotensin converting enzyme (kininase II). II. Immunocytochemistry and immunofluorescence, *Tissue Cell* 8:125–145, 1976.

Saijonmaa O, Nyman T, Fyhrquist F: Downregulation of angiotensin-converting enzyme by tumor necrosis factor-alpha and interleukin-1beta in cultured human endothelial cells, *J Vasc Res* 38:370–378, 2001.

Sakurai T, Okumura H, Matsumoto M, et al.: Endoglin (CD105) is a useful marker for evaluating microvessel density and predicting prognosis in esophageal squamous cell carcinoma, *Anticancer Res* 34:3431–3438, 2014.

Segarra M, Aburto MR, Cop F, et al.: Endothelial Dab1 signaling orchestrates neuro-glia-vessel communication in the central nervous system, *Science* 361:eaao2861, 2018.

Serluca FC, Drummond IA, Fishman MC: Endothelial signaling in kidney morphogenesis: a role for hemodynamic forces, *Curr Biol* 12:492–497, 2002.

Shah NM, Groves AK, Anderson DJ: Alternative neural crest cell ates are instructively promoted by TGFβ superfamily members, *Cell* 85:331–343, 1996.
Shen Q, Goderie SK, Jin L, et al.: Endothelial cells stimulate self-renewal and expand neurogenesis of neural stem cells, *Science* 304:1338–1340, 2004.
Smith JM, Meinkoth JH, Hochstatter T, et al.: Differential distribution of von Willebrand factor in canine vascular endothelium, *Am J Vet Res* 57:750–755, 1996.
Starke RD, Ferraro F, Paschalaki KE, et al.: Endothelial von Willebrand factor regulates angiogenesis, *Blood* 117:1071–1080, 2011.
Streeter PR, Rouse BT, Butcher EC: Immunohistologic and functional characterization of a vascular address in involved in lymphocyte homing into peripheral lymph nodes, *J Cell Biol* 107:1853–1862, 1988a.
Streeter PR, Berg EL, Rouse BTN, et al.: A tissue-specific endothelial cell molecule involved in lymphocyte homing, *Nature* 331:41–46, 1988b.
Tufro A, Norwood VF, Carey RM, et al.: Vascular endothelial growth factor induces nephrogenesis and vasculogenesis, *J Am Soc Nephrol* 10:2125–2134, 1999.
Turner RR, Beckstead JH, Warnke RA, et al.: Endothelial cell phenotypic diversity: in situ demonstration of immunologic and enzymatic heterogeneity that correlates with specific morphologic subtypes, *Am J Clin Pathol* 87:569–575, 1987.
Voermans V, Rood PML, Hordijk PL, et al.: Adhesion molecules involved in transendothelial migration of human hematopoietic progenitor cells, *Stem Cells* 18:435–443, 2000.
Wang B, Zhao L, Fish M, et al.: Self-renewing diploid Axin2(+) cells fuel homeostatic renewal of the liver, *Nature* 524:180–185, 2015.
Watt SM, Gschmeissner SE, Bates PA: PECAM-1: its expression and function as a cell adhesion molecule on hemopoietic and endothelial cells, *Leuk Lymphoma* 17:229–244, 1995.
Whitman MC, Fan W, Rela L, et al.: Blood vessels form a migratory scaffold in the rostral migratory stream, *J Comp Neurol* 516:94–104, 2009.
Yim SE, Shah Y, Tomita S, et al.: Disruption of the Arnt gene in endothelial cells causes hepatic vascular defects and partial embryonic lethality, *Hepatology* 44:550–560, 2006.
Yoshitomi H, Zaret KS: Endothelial cell interactions initiate dorsal pancreas developmentby selectively inducing the transcription factor Ptf1a, *Development* 131:807–817, 2004.
Zeng X, Wert SE, Federici R, et al.: VEGF enhances pulmonary vasculogenesis and disrupts lung morphogenesis, *Dev Dyn* 211:215–227, 1998.

CHAPTER 19
Concluding remarks

The generation of blood vessels was first described through careful observations of living amphibian or avian embryos under a microscope (Clark, 1918; Sabin, 1917, 1920), fixed and stained with classical dyes, and injected with India ink (Evans, 1909). These mainly descriptive studies established important basic concepts concerning the morphology and growth of blood vessels. Vascular morphogenesis is a complex process, which has implicated multiple factors at different levels of structural organization.

Modern experimental approaches can provide further insights into elucidating the mechanisms of angiogenesis and can also provide powerful tools for examining this process that governs the new formation of vessels under both physiological and pathological conditions. Among the in vitro models of sprouting angiogenesis, the sprouting model in which human umbilical vein endothelial cell (HUVEC)-coated beads in a fibrin matrix sprout and form lumenized vessels (Nakatsu et al., 2007) should be noted. Another model allows a programmed differentiation of mouse embryonic stem cells to form lumenized vessels (Kearney and Bautch, 2003). Embryoid bodies are clusters of embryonic stem cells that are able to differentiate into multiple cell lineages, including endothelial cells, which can be used to investigate quantitative and cell gene functions in sprouting angiogenesis (Jakobbson et al., 2010).

Computational modeling is a powerful predictive tool for understanding vascular development. In this context, in silico and in vitro studies highlight the importance of cell shape (elongated vs round) during vascular development (Merks et al., 2006). The initial sprouting process could be replicated in silico, which is an indication that vascular endothelial growth factor (VEGF)/Notch regulation is sufficient for patterning this process (Bentley et al., 2009). This model also predicted that the balance between VEGF receptor-1 and -2 (VEGFR-1 and VEGFR-2) expressions in endothelial cells affects their potential to become the tip cell in a growing sprout (Jakobbson et al., 2010). Another modeling approach that has been used to study cell behaviors and cell-cell communication in sprouting angiogenesis is the so-called "agent-based modeling" (ABM) (Qutub and Popel, 2009).

New Insights on the Development of the Vascular System
https://doi.org/10.1016/B978-0-323-90599-2.00019-2

Artel et al. (2011) developed an ABM that included sprout initiation, guidance, extension, and anastomosis to study the relationship between the pore size of an ex vivo scaffold and the rate of invasive angiogenesis. Arima et al. (2011), using time-lapse imaging in an aortic ring assay and a combination of labeling techniques that allows nuclear tracking of endothelial cells, performed a comprehensive analysis of cell movements and their relative contribution to sprout elongation. These models have provided new information about cell behaviors and their molecular regulation during blood vessel sprouting, even if a full understanding of how angiogenic sprouting and vessel network formation are regulated and several open questions regarding the spatial organization of signals and responses will require the integration of data from different model systems.

Although much knowledge of the molecular mechanisms of vasculogenesis and angiogenesis has been generated in the past few years, critical questions on how the vessels choose and follow specific paths to reach their targets remain unanswered. The classical literature provides a thorough coverage of the descriptive aspects of development of the embryonic vascular system. Modern techniques, such as the technology of cell fusion, cell sorting, and image analysis, can now provide insights into the mechanisms of these dynamics events during which vessels form and regress and blood flow changes directions in the same vessels, and can also provide powerful tools for examining the processes that govern the development of the vascular system.

Capillary endothelial cells are not merely passive conduits for the delivery of oxygen and nutrients but are also involved in organ morphogenesis and regeneration. Although microvascular endothelial cells were originally believed to form a rather inert structure, they are now considered to actively control the tissue microenvironment by endothelial cell-derived angiocrine factors, which control organogenesis during development and organ function in both health and disease.

Dynamic interactions between specific organ endothelial cells and different cellular components are crucial for different organ morphogenesis and functions. Conversely, organs provide cues, shaping vascular network structure. Paracrine interactions between vessels and developing organs play a crucial role, and relatively little is known about the molecules involved in these processes, even if recent advances in the understanding of the molecular mechanisms driving organotypic vascular differentiation have been provided (Augustin and Koh, 2017; Rafii et al., 2016).

Endothelial cell diversity has crucial implications in the development of vascular diseases. Systemic vasculitides and different septic reactions target different segments and branches of the vascular tree and selective vascular beds. Even thrombotic or hemorrhagic conditions recognize specific vascular beds at the sites of disease occurrence. Potential implications for the pathogenesis of vascular metabolic diseases are also strong. Finally, endothelial cell differences exist in the tumor vasculature at different stages.

Overall, blood vessels exert regulatory functions for the development of disease processes and are key targets for novel therapies for metabolic and inflammatory diseases as well as cancer.

References

Arima S, Nishiyama K, Ko T, et al.: Angiogenic morphogenesis driven by dynamic and heterogeneous collective endothelial cell movement, *Development* 138:4763–4776, 2011.

Artel A, Mehdizadeh H, Chiu YC, et al.: An agent-based model for the investigation of neovascularization within porous scaffolds, *Tissue Eng A* 17:2133–2141, 2011.

Augustin HG, Koh GY: Organotypic vasculature: from descriptive heterogeneity to functional pathophysiology, *Science* 357:eaal2379, 2017.

Bentley K, Mariggi G, Gerhardt H, et al.: Tipping the balance: robustness of tip selection, migration and fusion in angiogenesis, *PLoS Comput Biol* 5, 2009, e1000549.

Clark ER: Studies on the growth of blood vessels in the tail of the frog larva by observation and experiment on the living animal, *Am J Anat* 23:37–88, 1918.

Evans HM: On the development of the aorta, cardinal and umbilica veins, and other blood vessels of enbryos from capillaries, *Anat Rec* 3:498–518, 1909.

Jakobbson L, Franco CA, Bentley K, et al.: Endothelial cells dynamically compete for the tip cell position during angiogenic sprouting, *Nat Cell Biol* 12:943–953, 2010.

Kearney JB, Bautch VL: In vitro differentiation of mouse ES cells: hematopoietic and vascular development, *Methods Enzymol* 365:83–98, 2003.

Merks RM, Brodsky SV, Goligorsky MS, et al.: Cell elongation is key to in silico replication of in vitro vasculogenesis and subsequent remodeling, *Dev Biol* 289:44–54, 2006.

Nakatsu MN, Davis J, Hughes CC: Optimized fibrin gel bead assay from the study of angiogenesis, *J Vis Exp* 3:186, 2007.

Qutub AA, Popel AS: Elongation, proliferation and migration differentiate endothelial cell phenotypes and determine capillary sprouting, *BMC Syst Biol* 3:13, 2009.

Rafii S, Butler JM, Ding BS: Angiocrine functions of organ-specific endothelial cells, *Nature* 529:316–325, 2016.

Sabin F: Origin and development of the primitive vessels of the chick embryo and of the pig, *Contrib Embryol* 6:62–124, 1917.

Sabin F: Studies on the origin of blood vessels of the chick during the second day of incubation, *Contrib Embryol* 9:213–262, 1920.

Index

Note: Page numbers followed by *f* indicate figures.

D

E

F

G

H

I

J

K

L

M

N

O

P

X

Z

9780323905992